KELLER PLAN
for
SELF-PACED STUDY
using Masterton and Slowinski's
CHEMICAL PRINCIPLES

Second Edition

JOSEPH L. CLOUSER

Chairman, Department of Chemistry
William Rainey Harper College
Palatine, Illinois

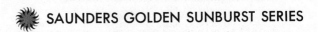 SAUNDERS GOLDEN SUNBURST SERIES

W. B. SAUNDERS COMPANY · Philadelphia · London · Toronto

W. B. Saunders Company: West Washington Square
Philadelphia, PA 19105

1 St. Anne's Road
Eastbourne, East Sussex BN21 3UN, England

1 Goldthorne Avenue
Toronto, Ontario M8Z 5T9, Canada

Keller Plan for Self-Paced Study Using
Masterton and Slowinski's *Chemical Principles*   ISBN 0-7216-2611-4

© 1977 by W. B. Saunders Company. Copyright 1974 by W. B. Saunders Company. Copyright under the International Copyright Union. All rights reserved. This book is protected by copyright. No part of it may be reproduced, stored in a retrieval system, or transmitted in any form or by any means, electronic, mechanical, photocopying, recording, or otherwise, without written permission from the publisher. Made in the United States of America. Press of W. B. Saunders Company. Library of Congress catalog card number 76-20081

Last digit is from the print number: 9   8   7   6   5   4   3

PREFACE

This Keller Plan Study Guide has been written as an aid to the student in his or her study of general chemistry using *Chemical Principles*, Fourth Edition, by W. L. Masterton and E. J. Slowinski.

Rather than providing objectives for an entire chapter, this Study Guide lists instructional objectives that are keyed to each section in the parent text. An objective for which a skill is to be developed is indicated by the * notation. These skills are illustrated in the unit summary and by reference to appropriate examples and problems in the text.

Each unit summary includes a brief review of important concepts presented in the text. Sections in the summary contain additional exercises with detailed solutions. The summaries of some of the early units contain a somewhat detailed discussion of exponential notation, significant figures, problem solving using the conversion factor approach, and the mole concept.

Following the unit summary, a suggested assignment is given. This assignment includes problems from the answered member of the "matched pair" at the end of the chapter in the parent text. Each unit concludes with worked-out solutions to the assigned problems.

This Guide is intended to provide material for study and mastery of chemical principles, concepts, and applications for self-paced, independent study in a course using Masterton and Slowinski's *Chemical Principles*, Fourth Edition.

I wish to acknowledge and thank Professor William L. Masterton for his assistance in the preparation of the manuscript for this book. Comments and suggestions from students and instructors who used the first edition have been very helpful. I appreciate the encouragement and support of the W. B. Saunders Company, particularly that of Mr. John Vondeling. Finally, I would like to thank Mary for typing the manuscript.

JOSEPH L. CLOUSER

TO THE STUDENT

The format of a Keller Plan course will vary somewhat from the usual lecture-discussion type of course. The Keller Plan is designed to allow you to proceed, within limits, at your own pace, mastering principles in one unit before proceeding to the next. The key to success in this course is to become actively involved. You must become a participant instead of passively receiving the information that is conveyed in a lecture. This is not a "spectator" course but one that requires your active involvement.

How to Use this Guide

The book is divided into units that correspond to chapters in your text, *Chemical Principles*, Fourth Edition, by Masterton and Slowinski. Each unit contains a list of instructional objectives, a unit summary, suggested assignment, and solutions to assigned problems. The following procedures are suggested for each unit.

1. Read the goals and objectives for the unit as listed in this guide. The objectives are keyed to each of the sections in *Chemical Principles*. The objectives serve to focus your attention and indicate to you the important concepts and applications developed in that section. The objectives are both qualitative and quantitative. The objectives marked with an asterisk are usually quantitative and require mathematical skill. These skills are illustrated by exercises in the summary of this guide and by reference to specific problems in the text.

2. With the objectives in mind and at hand for ready reference, read and study the corresponding section in the text.

3. Now read the unit summary. The summary serves to emphasize the concepts and mathematical applications that are developed in the text. Additional explanations and sample problems are included here. Work each problem and then consult the detailed solution. Use the summary and text together to be certain that you have mastered the material.

4. As a minimum, work out the suggested problems in detail, using units.

5. After making your best effort on the problems, consult the solutions. You need not solve a problem exactly as shown in the solutions, but you should feel that you thoroughly understand your method of solution. Be sure that you use proper units throughout the solution. Express your answer with the correct number of significant figures.

6. After reviewing the solutions to the assigned problems, see the instructor for help with any of the problems that you do not understand. For the problems that give you trouble, work the corresponding problem in the left column or additional problems from the right column for which answers are provided in the appendix of the text. Notice that problems at the end of the chapter are paired. For example, Problem 1.25, which is assigned, corresponds to problem 1.11 located directly across from it. Both problems illustrate the same principle and require development of the same type of skill.

7. Review the goals and objectives listed in the guide, and read the unit summary again. If you do not feel confident of your comprehension of the unit, consult your instructor or tutor for additional help.

8. Take the unit quiz. Be sure to discuss any errors with the proctor.

Additional Comments

Set work schedules and deadlines for yourself. You will find that some units will not require as much time and effort as others. Since you have a certain amount of freedom to set your own pace, allow additional time for units that you find difficult, and move at a more rapid rate through units that are easier for you. Your instructor will announce a grading policy, probably set some guidelines for completion of certain units, indicate the degree of mastery that is required on unit quizzes before you can proceed to the next unit, and set requirements for lectures. **Responsibility for success in this course rests with you.** You are graded on your individual performance and not on a class curve. If you develop self-discipline, don't procrastinate, and budget your time, you can employ this system for successful mastery of the principles covered in the course.

Additional Study Aids

Boyington, R., and W. L. Masterton, *Student's Guide to Masterton and Slowinski's Chemical Principles*, W. B. Saunders Company, 1973.

This guide is also keyed to the parent text. It contains questions to guide your study, chapter summaries, discussion of skills with examples and problems, and chapter self-tests consisting of true-false, and multiple choice questions and problems.

Peters, E. I., *Problem Solving for Chemistry*, 2nd ed., W. B. Saunders Company, 1976.

This is an auto-tutorial style text in which the student solves example problems through a series of steps. There are many problems, with complete solutions for about one-half the problems.

Masterton, W. L. and E. J. Slowinski, *Elementary Mathematical Preparation for General Chemistry*, W. B. Saunders Company, 1974.

This contains an extensive discussion of the mathematics essential for solving problems in general chemistry. It includes material on "setting up" problems, and contains many examples and problems with solutions.

CONTENTS

UNIT 1
Chemistry: An Experimental Science 1

UNIT 2
Atoms, Molecules, and Ions 7

UNIT 3
Chemical Formulas and Equations 19

UNIT 4
Thermochemistry 35

UNIT 5
The Physical Behavior of Gases 47

UNIT 6
The Electronic Structure of Atoms 59

UNIT 7
Periodic Table 69

UNIT 8
Chemical Bonding 77

UNIT 9
Physical Properties as Related to Structure 89

UNIT 10
Introduction to Organic Chemistry 97

UNIT 11
Liquids and Solids, Phase Changes 109

UNIT 12
Solutions 119

UNIT 13
Water, Pure and Otherwise 133

UNIT 14
Spontaneity of Reaction; ΔG and ΔS 141

UNIT 15
Chemical Equilibrium in Gaseous Systems 153

UNIT 16
Rates of Reaction 169

UNIT 17
The Atmosphere 185

UNIT 18
Precipitation Reactions 195

UNIT 19
Acids and Bases 211

UNIT 20
Acid-Base Reactions 227

UNIT 21
Complex Ions: Coordination Compounds 245

UNIT 22
Oxidation and Reduction: Electrochemical Cells 257

UNIT 23
Oxidation-Reduction Reaction: Spontaneity and Extent 271

UNIT 24
Nuclear Reactions 285

UNIT 25
Polymers: Natural and Synthetic 295

UNIT 1

Chemistry: An Experimental Science

The goal of this unit is to recognize the role that an experimental science such as chemistry plays in our lives. In addition, the principles of measurement and methods of separation and identification of pure substances will be considered.

INSTRUCTIONAL OBJECTIVES

Introduction
  Realize that principles developed from experiments on simple systems may lead to a variety of applications in our complex world.

1.1 Measurements
  1. Define the basic units of length, volume, mass, temperature, time, pressure, and energy.
  2. Describe instruments that may be used to measure mass and volume and their method of operation.
  3. Distinguish between mass and weight.
  4. Distinguish between heat and temperature and describe the experimental basis for the measurement of temperature.
  *5. Perform calculations to make conversions between the various temperature scales. (Prob. 1.3)
  *6. Express the uncertainty in a measurement, or in a calculation based upon measurements, by applying the rules of significant figures. (Probs. 1.1, 1.23 and all other problems)
  *7. Convert units within the metric system and between the metric

---

The notation * indicates an objective, usually quantitative in nature, for which a skill is to be developed. These skills are illustrated in the summary or by reference to appropriate examples and problems in the text.

and English systems through the use of the conversion factor method. (Probs. 1.2, 1.25, 1.28)
*8. Express numbers in exponential notation. (Appendix 4)
*9. Perform multiplication, division, addition, and subtraction operations using exponential notation. (Appendix 4) (Probs 1.22, 1.28)

1.2 Kinds of Substances
1. Define the terms: element, compound.
2. Become familiar with the names and symbols of some of the common elements. (Table 1.4)

1.3 Identification of Pure Substances
1. Distinguish between chemical and physical properties.
2. Describe methods used to identify and check the purity of a substance.
*3. Employ mass, volume, and density relationships in calculations. (Probs. 1.1, 1.22)

1.4 Separation of Matter into Pure Substances
1. Distinguish between homogeneous and heterogeneous mixtures.
2. Describe methods (and the principles upon which these methods are based) for the separation of pure substances from mixtures.

SUGGESTED ASSIGNMENT I

Text: Chapter 1, Appendix 4

Problems: Chapter 1, Numbers 1.1–1.3, 1.22, 1.23, 1.25, 1.28, 1.31

Solutions to Assigned Problems Set I

1.1

$$\text{Density} = \frac{\text{mass}}{\text{volume}}; \quad \text{Volume} = \frac{\text{mass}}{\text{density}} = \frac{35g}{2.70g/cm^3} = 13 \text{ cm}^3$$

Substituting into a defining equation is one approach to the solution of this problem. Notice that each number has a dimension, and that the unit cm³ results.

According to the rules governing the use of significant figures, when two measured values are multiplied or divided, the number of significant figures in the result cannot exceed that in the least precise measurement. In 35g and 2.70g/cm³, there are two and three significant figures, respectively. Therefore, a calculated result such as 12.962962 cm³ must be rounded to an answer that contains two significant figures. The correct answer, then, is 13 cm³.

1.2

We make use of the conversion factors from Table 1.1 as follows:

$$17.8 \frac{\text{miles}}{\text{gal}} \times \frac{1.609 \text{ km}}{\text{mile}} \times \frac{1 \text{ gal}}{4 \text{ qt}} \times \frac{1.057 \text{ qt}}{\text{liter}} = 7.57 \text{ km/liter}$$

Note that the result must be rounded to three significant figures.

1.3

Using Equation 1.3, which relates °K and °C,

°C = °K - 273 = 209 - 273 = -64°C

Applying Equation 1.2:

°F = 1.8°C + 32 = 1.8 (-64) + 32 = -83°F

1.22

a. Assume that the atom is spherical and calculate the volume of a sphere that has a radius of 1.26 Å.

First convert Å to cm. $1.26 \text{ Å} \times \dfrac{1 \times 10^{-8} \text{cm}}{\text{Å}} = 1.26 \times 10^{-8} \text{ cm}$

$$V = \dfrac{4\pi r^3}{3} = \dfrac{(4)(3.14)(1.26 \times 10^{-8} \text{ cm})^3}{3} = 8.37 \times 10^{-24} \text{ cm}^3$$

b. We need to convert amu to grams and divide by the volume of the iron atom.

$$55.8 \text{ amu} \times \dfrac{1.67 \times 10^{-24} \text{ g}}{1 \text{ amu}} \times \dfrac{1}{8.37 \times 10^{-24} \text{ cm}^3} = 11.1 \text{ g/cm}^3$$

c. There must be empty space between iron atoms in the bulk sample.

1.23

The value 85.279g is discarded since it is not consistent with the first three measurements.

Mass water = (mass of beaker + water) − mass of beaker

Trial 1: 84.136g − 74.242g = 9.894g
2: 84.151g − 74.242g = 9.909g
3: 84.141g − 74.242g = 9.899g

$$\text{Average mass water} = \dfrac{9.894\text{g} + 9.909\text{g} + 9.899\text{g}}{3} = 9.901\text{g}$$

$$\text{Volume} = \dfrac{\text{mass}}{\text{density}} = \dfrac{9.901\text{g}}{0.9970 \text{ g/cm}^3} = 9.931 \text{ cm}^3$$

1.25

Use Tables 1.1 and 1.2 for conversion factors.

a. $\dfrac{36.5\text{g}}{100\text{g}} \times \dfrac{1 \times 10^3 \text{g}}{\text{kg}} \times \dfrac{1 \text{ lb}}{453.6\text{g}} = 0.805 \text{ lb/kg water}$

b. $1.68 \text{ gal} \times \dfrac{4 \text{ qt}}{\text{gal}} \times \dfrac{1 \text{ liter}}{1.057 \text{ qt}} \times \dfrac{1 \text{ m}^3}{10^3 \text{ liter}} = 6.36 \times 10^{-3} \text{ m}^3$

c. $12.4 \text{ kcal} \times \dfrac{1 \times 10^3 \text{ cal}}{\text{kcal}} \times \dfrac{4.184 \text{ joules}}{\text{cal}} = 5.19 \times 10^4 \text{ joules}$

1.28

This is another problem involving conversion factors from Table 1.1.

$2.00 \text{ lb fish} \times \dfrac{453.6 \text{ g}}{1 \text{ lb}} \times \dfrac{0.50 \text{g Hg}}{10^6 \text{g fish}} = 4.5 \times 10^{-4} \text{ g Hg}$

1.31

Refer to Table 1.7 for the solubilities of these two acids.

At 20°C, the solubility of succinic acid is 7g/100g water.

50g − 7g = 43g of succinic acid crystallize from solution.

At 20°C, the solubility of tartaric acid is 18g/100g water.

50g − 18g = 32g of tartaric acid crystallize from solution.

UNIT 2

Atoms, Molecules, and Ions

The goal of this unit is to examine fundamental units of matter and to relate the properties of substances to these particles.

INSTRUCTIONAL OBJECTIVES

2.1 Atomic Theory
1. State the postulates of Dalton's atomic theory.
2. State the Laws of Conservation of Mass, Constant Composition, and Multiple Proportions.
3. Relate the Postulates of Dalton's theory to these laws.
*4. Illustrate the Law of Multiple Proportions when given weight analysis data for two or more compounds of two elements. (Prob. 2.1)

2.2 Components of the Atom
1. Summarize the experimental work of Thomson, Milliken, and Rutherford and state the conclusions resulting from their work.
2. List the charges and relative masses of a neutron, a proton, and an electron.
3. Recognize the existence of isotopes and the nuclear differences between the isotopes of an element.
*4. Illustrate the relationship of atomic number and mass number to the number of protons, neutrons, and electrons. (Prob. 2.2)

2.3 Molecules and Ions
1. Distinguish between atom, molecule, and ion.
2. Explain how cations and anions are formed from atoms.

2.4 Relative Masses of Atoms
1. Define atomic weight and explain its meaning.
2. Realize that the atomic weights are based on the carbon twelve scale.
3. Define gram atomic weight and explain its meaning.
4. Describe the basic principles of the mass spectrometer.
*5. Calculate the average atomic weight of an element when given the masses and abundances of each component isotope. (Prob. 2.4)
*6. Calculate the percentage abundance of each isotope when given the average atomic weight of an element and the masses of each constitutent isotope. (Prob. 2.30)

2.5 Masses of Atoms, Avogadro's Number
1. Recognize that Avogadro's number is the number of atoms in one gram atomic weight of an element.
*2. Perform the following:
   a. Calculate the mass of a sample when given the atomic weight of the element and the number of atoms. (Probs. 2.5a, 2.32b)
   b. Calculate the number of atoms of an element when given the mass of the sample. (Probs. 2.5b, 2.32a)

2.6 Masses of Molecules
1. Define molecular weight and explain its meaning.
2. Recognize that the gram molecular weight of a substance is the weight in grams which contains Avogadro's number of molecules.
*3. Perform the following:
   a. Calculate the molecular weight when given the composition of a molecule. (Prob. 2.6a)
   b. Calculate the mass of a sample of molecules when given the molecular weight and the number of molecules. (Prob. 2.33e)
   c. Calculate the number of molecules when given the mass of a sample. (Prob. 2.6b)

SUMMARY

1. Laws of Conservation of Mass, Constant Composition, and Multiple Proportions

These laws are the foundation for many quantitative relationships governing chemical composition and chemical reactions. The Law of Conservation of Mass is used to determine the number of grams of oxygen that combines with aluminum in this example:

If 1.000g of aluminum oxide is formed when 0.529g of aluminum is burned in an open dish, then (1.000 − 0.529)g or 0.471g of oxygen must have chemically combined with the aluminum. The mass of oxygen that reacted was determined from the relationship: mass of aluminum reacted + mass of oxygen reacted = mass of aluminum oxide produced. This relationship is based on the law that states that there is no detectable change in mass in an ordinary chemical reaction. If the masses of all but one of the reactants or products are known, the unknown mass can be calculated by applying the Law of Conservation of Mass. Consider the same aluminum-oxygen reaction but this time react 2.645g of aluminum to produce 5.000g of aluminum oxide. The mass of oxygen that reacted was (5.000 − 2.645)g or 2.355g.

Let us consider the data from the two reactions above from the point of view of the composition of the product. In reaction one: per cent aluminum in the product =

$$\frac{\text{wt. aluminum}}{\text{wt. aluminum oxide}} \times 100 = \frac{0.529g}{1.000g} \times 100 = 52.9\% \text{ Al}$$

$$\text{per cent oxygen in the product} = \frac{\text{wt. oxygen}}{\text{wt. aluminum oxide}} \times 100$$

$$= \frac{0.471g}{1.000g} \times 100 = 47.1\% \text{ O}$$

Similar calculations for reaction two are:

$$\frac{2.645g}{5.000g} \times 100 = 52.9\% \text{ Al}; \frac{2.355g}{5.000g} \times 100 = 47.1\% \text{ O}$$

These calculations illustrate the Law of Constant Composition. A compound always contains the same elements in the same proportions by weight. No matter where or how aluminum oxide is prepared, it will always be composed of 52.9% Al and 47.1% O by weight.

Some elements form more than one compound with oxygen; for example, carbon reacts to form carbon dioxide under certain conditions of reaction and carbon monoxide under other conditions. Copper forms two different oxides as well as two different sets of chlorides, bromides, and sulfides. It is found that when two elements combine to form more than one compound, the masses of one element that combine with a fixed mass of the other element are in the ratio of small whole numbers. This is a statement of the Law of Multiple Proportions.

The behavior of copper and bromine illustrates this law. One compound formed from copper and bromine is white in color, while another compound formed from the same elements is black. Analysis of the white colored compound shows the composition to be 44.29% copper and 55.71% bromine. The black compound is 28.45% copper and 71.55% bromine. (Remember that percentage composition means that if we had one hundred grams of the white colored compound, we would have 44.29g copper and 55.71g bromine.)

If we select copper to be the fixed mass, for the white compound there is: $\frac{55.71\text{g bromine}}{44.29\text{g copper}} = 1.258 \frac{\text{g bromine}}{\text{g copper}}$. For the black compound: $\frac{71.55\text{g bromine}}{28.45\text{g copper}} = 2.515 \frac{\text{g bromine}}{\text{g copper}}$.

Note that we have calculated the mass of bromine that combines with a fixed mass (one gram) of copper. The ratio of the masses of bromine per gram of copper in the two compounds is $\frac{2.515\text{g}}{1.258\text{g}} = 1.999 \approx 2$, a small whole number. If bromine is selected to have the fixed mass, then the white compound has $0.7950 \frac{\text{g copper}}{\text{g bromine}}$ and the black compound $0.3976 \frac{\text{g copper}}{\text{g bromine}}$. The ratio of the masses of copper per gram of bromine in the white compound to that in the black compound is $\frac{0.7950\text{g}}{0.3976\text{g}} = 1.999 \approx 2$, a small whole number.

Exercise 2.1
1. Exactly 2.000g of lead reacted with sulfur to produce 2.310g of lead sulfide. Calculate the percentage of sulfur in the lead sulfide.
2. Exactly 1.598g of copper is heated at low temperature to produce 2.000g of a black-colored copper-oxygen compound. When this black-colored compound is heated to a high temperature, a red-colored copper-oxygen compound weighing 1.799g is produced and oxygen gas is released. Show how these data illustrate the Law of Multiple Proportions.

Exercise 2.1 Answers
1. First use the Law of Conservation of Matter to determine the mass of sulfur that combined.

    2.310g − 2.000g = 0.310g = g sulfur that reacted

    % sulfur = $\frac{\text{wt. sulfur}}{\text{wt. lead sulfide}} \times 100 = \frac{0.310\text{g}}{2.310\text{g}} \times 100 = 13.4\%$ S
2. Before we can calculate the mass of copper that combines with one gram of oxygen, we must use the Law of Conservation of Mass to calculate the number of grams of oxygen that combined with 1.598g of copper to form the black-colored compound. 2.000g − 1.598g = 0.402g oxygen.

    A similar calculation gives the number of grams of oxygen in 1.799g of the red-colored copper-oxygen compound.

    1.799g − 1.598g = 0.201g oxygen

For the black compound: $\dfrac{1.598 \text{g copper}}{0.402 \text{g oxygen}} = 3.98 \dfrac{\text{g copper}}{\text{g oxygen}}$

For the red compound: $\dfrac{1.598 \text{g copper}}{0.201 \text{g oxygen}} = 7.95 \dfrac{\text{g copper}}{\text{g oxygen}}$

The ratio of the masses of copper per gram of oxygen in the two compounds is $\dfrac{7.95 \text{g copper}}{3.98 \text{g copper}} = 1.99$ or $2:1$.

2. Electrons, Protons, and Neutrons

Thomson's experiments using a cathode-ray tube demonstrated the existence of fundamental particles called electrons. The mass of the electron could be determined when Milliken's calculation of the charge on an electron was combined with Thomson's mass-to-charge ratio. As a result of alpha-particle scattering by thin metal foils, Rutherford concluded that the atom must have a very small positively charged nucleus that contained almost all of the mass of the atom. The nucleus is thought to contain protons and neutrons. The proton has a mass nearly equal to the mass of a hydrogen atom and is assigned a charge of +1 while the electron is assigned a charge of −1. Note that the assigned charge of −1 is really a relative charge and corresponds to an actual charge of 1.60×10^{-19} coulombs. The neutron is an uncharged particle with a mass about equal to the mass of a proton.

The number of electrons equals the number of protons in a neutral atom and the number of protons is equal to the atomic number. Isotopes of the same element have the same atomic number but different numbers of neutrons. The mass number is equal to the sum of the number of protons and the number of neutrons. The nuclear symbol indicates the atomic number as a subscript at the lower left of the symbol of the element and the mass number as a superscript at the upper left.

A molecule is an aggregate of atoms held together by chemical bonds. Ions are charged particles as opposed to atoms and molecules, which are electrically neutral. Cations are positively charged ions that are formed by the loss of electrons from neutral species. Anions are negatively charged ions that are formed when a neutral species gains electrons.

Exercise 2.2

Complete the following table:

Species	Number of Protons	Number of Neutrons	Number of Electrons	Charge
A. $^{9}_{4}\text{Be}$				0
B. $^{52}_{24}\text{Cr}$				+3

Species	Number of Protons	Number of Neutrons	Number of Electrons	Charge
C.	9	10	10	
D.	17	18		-1

Exercise 2.2 Answers

A. Four protons and 4 electrons, the atomic number is 4. There are 5 neutrons, the difference between a mass number of 9 and the atomic number 4. There is zero net charge.

B. Twenty-four protons (the atomic number is 24). The mass number minus the atomic number is equal to the number of neutrons, $52 - 24 = 28$. Since the charge is +3, there must be three less electrons than protons, $24 - 3 = 21$.

C. The number of protons (9) is equal to the atomic number. The sum of the number of protons and the number of neutrons is equal to the mass number, $9 + 10 = 19$. Since there is an excess of one electron, the species has a net -1 charge. The species is $^{19}_{9}F^{-}$. Note that you need to refer to a periodic table to determine the symbol for the element of atomic number 9.

D. Seventeen protons and 18 neutrons indicate that the atomic number is 17 and the mass number is 35. If the species has a net charge of -1, there must be one more electron than protons or 18 electrons. The proper nuclear symbol is $^{35}_{17}Cl^{-}$.

3. Atomic Weights, Gram Atomic Weights and the Masses of Atoms

The atomic weight of an element is a number that indicates how heavy, on the average, an atom of an element is compared to an atom of another element. If a collection of atoms of one element contains the same number of atoms as another element, then the ratio of weights of the samples will be the same as the ratio of weights of the individual atoms. The standard of comparison now in use is based on carbon-12. The number of atoms in twelve grams of carbon-12 is 6.02×10^{23}, Avogadro's number. If we have 6.02×10^{23} atoms, we have one gram atomic weight of the element and a mass in grams equal to the atomic weight of the element.

The mass spectrometer is an instrument used to measure the mass-to-charge ratio of positively charged particles. Atomic weights can be determined from the voltage or field strength required to bring two ions of the same charge to the same point in the mass spectrometer. When more than one isotope of an element is present, the mass and relative abundance of each isotope must be known to calculate the average atomic weight.

The following exercise illustrates the principles described in the above summary.

Exercise 2.3
1. Calculate the weight in grams of one copper atom.
2. Calculate the number of copper atoms in a sample weighing 1.000g.
3. Copper consists of two isotopes, ^{63}Cu (mass = 62.929), and ^{65}Cu (mass = 64.928). Calculate the percentage abundance of the ^{63}Cu isotope.
4. Naturally occuring antimony contains 57.25% ^{121}Sb with a mass of 120.904 and 42.75% ^{123}Sb with a mass of 122.904. Calculate the average atomic weight of antimony.

Exercise 2.3 Answers

1. $1 \text{ atom} \times \dfrac{63.54 \text{g}}{6.02 \times 10^{23} \text{ atoms}} = 1.06 \times 10^{-22} \text{g}$

2. $1.000 \text{g} \times \dfrac{6.02 \times 10^{23} \text{ atoms}}{63.54 \text{g}} = 9.47 \times 10^{21} \text{ atoms}$

Notice that these solutions illustrate the conversion factor approach that was discussed in Chapter 1. If you did not use units in these solutions, review the material on problem solving in the preceding chapter.

3. Let x equal the fraction of the ^{63}Cu isotope. The fraction of the ^{65}Cu must be (1 - x). The equation is:

$x(62.929) + (1 - x)(64.928) = 63.54$

$62.929x + 64.928 - 64.928x = 63.54$

$-1.999x = -1.388; x = 0.6943$

Converting to percentage, the abundance of ^{66}Cu is 69.43%.

4. Multiply the mass of each isotope by its fractional abundance and then add.

$0.5725 \times 120.904 \quad = \quad 69.22$

$0.4275 \times 122.904 \quad = \quad \underline{52.54}$

atomic weight of antimony = 121.76

4. Molecular Weights, Gram Molecular Weights, and the Masses of Molecules

The molecular weight is a number that indicates how heavy a molecule is as compared to an atom of carbon-12. The molecular weight is the sum of the atomic weights of all the atoms in the molecule. The number of grams numerically equal to the molecular weight is the gram molecular weight and contains 6.02×10^{23} molecules. The following exercise illustrates the relationships among these quantities.

Exercise 2.4
1. Calculate the molecular weight of carbon dioxide; each molecule contains one atom of carbon and two atoms of oxygen.
2. Calculate the number of molecules in 16.0g of carbon dioxide.
3. Calculate the weight in grams of one molecule of carbon dioxide.

Exercise 2.4 Answers
1. MW = AW carbon + 2 (AW oxygen) = 12.01 + 2(16.00) = 44.01
2. $16.0\text{g} \times \dfrac{6.02 \times 10^{23}\,\text{molecules}}{44.01\text{g}} = 2.19 \times 10^{23}$ molecules
3. $1\text{ molecule} \times \dfrac{44.01\text{g}}{6.02 \times 10^{23}\,\text{molecules}} = 7.31 \times 10^{-23}\text{g}$

SUGGESTED ASSIGNMENT II

Text: Chapter 2

Problems: Chapter 2, Numbers 2.1–2.5, 2.6a, b, 2.30, 2.32a, b, 2.33

Solutions to Assigned Problems Set II

2.1

a. One approach to percentage composition problems is to assume one hundred grams of compound. In 100 grams of compound (1) there would be 79.89g Cu and 20.11g O. In compound (2), 88.82g Cu and 11.18g O. The weight of copper per gram of oxygen in each compound is:

$$\frac{79.89 \text{gCu}}{20.11 \text{gO}} = 3.973 \text{gCu/gO}; \frac{88.82 \text{gCu}}{11.18 \text{gO}} = 7.945 \text{gCu/gO}$$

Note that four significant figures are justified, and that the units are in grams copper per gram oxygen.

b. To illustrate the Law of Multiple Proportions, we need to compare the weights of copper in each of the two compounds that combine with a fixed weight (one gram in this case) of oxygen.

$$\frac{7.945 \text{ gCu (compound 2)}}{3.973 \text{ gCu (compound 1)}} = \frac{2.000}{1.000}$$

Clearly, the masses are in a ratio of small whole number, 2:1.

2.2

a. The atomic number (9) is equal to the number of protons. In an electrically neutral atom, the number of electrons is equal to the number of protons. Therefore, there are nine electrons. The mass number is equal to the number of protons plus the number of neutrons. The number of neutrons = 19 - 9 = 10.

b. The mass number is 50 + 71 = 121. The atomic number is equal to the number of protons. The symbol is $^{121}_{50}$Sn.

2.3

a. The ratio of the masses of two atoms is the same as the ratio of their atomic weights. For arsenic and oxygen, $\frac{74.9}{16.0} = 4.68$. This means that an arsenic atom weighs 4.68 times as much as an oxygen atom.

b. The atomic weight of aresenic must be 4.68 times that of oxygen as indicated in (a). If the atomic weight of oxygen were 100, the atomic weight of arsenic would be $4.68 \times 100 = 468$.

2.4

The average atomic weight can be obtained by adding the contributions of each of the isotopes. The fraction of the isotope with a mass of 10.02 is 0.1883. The fraction of the isotope with a mass of 11.01 is 0.8117.

Average atomic weight B = (10.02)(0.1883) + (11.01)(0.8117)

$$= 1.887 + 8.937 = 10.82$$

2.5

a. One gram atomic weight of Cd weighs 112g and is the weight of 6.02×10^{23} atoms.

$112g/6.02 \times 10^{23}$ atoms $= 1.86 \times 10^{-22}$ g/atom

Note that the answer can contain three significant figures, and that by labeling the numbers, the units of the answer indicate the weight of one atom. If we had set up the problem as: 6.02×10^{23} atoms/112g we would have obtained the number of atoms per one gram Cd. Obviously, that is not what was called for in the problem.

b. Again, we know that one gram atomic weight of Cd weighs 112g. Using this as a conversion factor, and labeling all numbers

$$28.0g \times \frac{1 \text{ GAW Cd}}{112g} = 0.250 \text{ GAW Cd}$$

The correct answer contains three significant figures.

2.6

a. Recall that the molecular weight is the sum of the atomic weights of the atoms in the molecule. The GMW is the weight in grams of Avogadro's number of molecules.

MW acetylene = 2 (AW carbon) + 2 (AW hydrogen)

$$= 2 (12.01) + 2 (1.008)$$

$$= 24.02 + 2.016 = 26.04$$

The GMW of acetylene is 26.04g.

b. Knowing that 26.04g is the weight of 6.02×10^{23} molecules, we use this as a conversion factor

$$13.0g \times \frac{6.02 \times 10^{23} \text{ molecules}}{26.04g} = 3.01 \times 10^{23} \text{ molecules}$$

2.30

The isotope of mass 27.985 has a 92.28% abundance. The other two isotopes must have (100.00 - 92.28)% = 7.72% abundances; 0.0772 represents the total fraction for both. Let x = fraction of the isotope with a mass of 28.99. Then $0.0772 - x$ = fraction of the isotope wiht a mass of 29.98. The average atomic weight is 28.086.

$$28.086 = 27.985 (0.9228) + (28.99) (x) + (29.98)(0.0772 - x)$$

$$= 25.825 + 28.99x + 2.314 - 29.98x$$

$$x = 0.0535$$

It has 5.35% abundance for the isotope of mass 28.99, and 2.37% abundance for the isotope of mass 29.98.

2.32

a. Use the conversion factor 6.02×10^{23} atoms = 28.1g Si

$$1.43g \times \frac{6.02 \times 10^{23} \text{ atoms}}{28.1g} = 3.06 \times 10^{22} \text{ atoms}$$

b. One gram atomic weight of Si weighs 28.1g and is the weight of 6.02×10^{23} atoms.

$$28.1g / 6.02 \times 10^{23} \text{ atoms} = 4.67 \times 10^{-23} \text{ g/atom}$$

2.33

Calculate the mass of each and arrange in order of increasing mass.

a. $63.5\text{g}/6.02 \times 10^{23}\text{ atoms} = 1.05 \times 10^{-22}\text{ g}$
b. $1 \times 10^{-21}\text{ GAW} \times 63.5\text{g/GAW} = 6 \times 10^{-20}\text{ g}$
c. $1 \times 10^{-21}\text{ GMW} \times 18.0\text{g/GMW} = 1.8 \times 10^{-20}\text{ g}$
d. The Cu^{2+} would have a very slightly smaller mass than a Cu atom.
e. 6 molecules $\times 18.0\text{g}/6.0 \times 10^{23}$ molecules $= 1.8 \times 10^{-22}\text{ g}$

In order of increasing mass: $d < a < e < c < b$

UNIT 3

Chemical Formulas and Equations

The goal of this unit is to represent the composition of substances with chemical formulas and to write and interpret chemical equations.

INSTRUCTIONAL OBJECTIVES

3.1 Chemical Formulas
1. Review the meaning of atomic weight, gram atomic weight, molecule, ion, molecular weight, and gram molecular weight.
2. Distinguish between empirical and molecular formulas.
*3. State the number of atoms of each element in a molecule when given the formula of a molecular species.
*4. Write the molecular formula when given the number of atoms of each element.

3.2 Simplest Formulas from Analysis
*1. Determine the empirical formula of a compound when given percentages by weight of the elements or analytical data from which relative weights of the elements may be calculated. (Probs. 3.1, 3.30)
*2. Calculate the percentage composition when given the chemical formula of a compound. (Prob. 3.4)

3.3 Molecular Formula from Simplest Formula
*1. Determine the molecular formula when given the empirical formula, or data from which the empirical formula can be determined, and the molecular weight, or information from which the molecular weight can be calculated. (Prob. 3.2)

3.4 The Mole
1. Realize that the mole contains Avogardro's number of formula units.
2. Remember that the mole is always associated with a formula.
3. Remember that a mole represents a definite mass that is associated with this number of particles.
*4. Make conversions between the number of moles, number of grams, and the number of particles (atoms, molecules, or formula units). (Probs. 3.3, 3.34)

3.5 Chemical Equations
*1. Write and balance equations when given the formulas of the reactants and products. (Probs. 3.5, 3.36)

3.6 Mass Relations in Reactions
1. Recognize that the coefficients of a balanced equation indicate the relative number of moles of reactants and products.
*2. Use conversion factors to relate the numbers of moles, grams, and particles when given a balanced equation. (Probs. 3.6, 3.43)

3.7 Limiting Reagent, Theoretical and Actual Yields
1. Distinguish between theoretical and actual yields.
*2. Calculate which reactant is the limiting reagent when given a balanced equation and the quantity of each reactant. (Prob. 3.7a)
*3. Calculate the theoretical yield of product based on the limiting reagent. (Prob. 3.7b)
*4. Determine the percentage yield when given the actual yield of product and the theoretical yield or data from which the theoretical yield can be calculated. (Probs. 3.7c, 3.43)

SUMMARY

1. The Mole

It is very useful to think of a mole as the Avogadro number of objects, just as a dozen indicates twelve objects. Thus we can indicate a mole of electrons, a mole of H atoms, a mole of H_2 molecules, or a mole of dollars. In each case we are indicating 6.02×10^{23} objects. However, it is very important that the nature of the objects be clearly specified. If we say that we have a mole of hydrogen, it is not clear whether we are referring to hydrogen atoms or hydrogen molecules. To avoid ambiguity, be certain to specify exactly what objects you are counting.

Exercise 3.1

Use conversion factors to solve the following:

1. A sample of carbon tetrachloride, CCl_4, contains 4.1×10^{23} molecules. Calculate the number of moles of CCl_4 present.
2. Calculate the number of P_4 molecules in a 0.65 mole sample of P_4.
3. Calculate the number of P atoms in a 0.65 mole sample of P_4 molecules.
4. Calculate the number of moles of P atoms in a 0.65 mole sample of P_4 molecules.

Exercise 3.1 Answers

1. 4.1×10^{23} molecules $\times \dfrac{1 \text{ mole}}{6.02 \times 10^{23} \text{ molecules}} = 6.8 \times 10^{-1}$ mole.

   Note that the answer can contain only two significant figures.

2. 0.65 mole $P_4 \times \dfrac{6.02 \times 10^{23} \text{ molecules}}{\text{mole } P_4} = 3.9 \times 10^{23}$ molecules

3. One P_4 molecule contains four P atoms.

   $3.9 \times 10^{23}\, P_4$ molecules $\times \dfrac{4 \text{ P atoms}}{1\, P_4 \text{ molecule}} = 1.6 \times 10^{24}$ P atoms

4. 0.65 moles P_4 molecules $\times \dfrac{4 \text{ moles P atoms}}{1 \text{ mole } P_4 \text{ molecules}}$

   = 2.6 moles P atoms

The mole can also be considered to be the mass in grams represented by one formula weight of a substance. Knowing the formula of a substance and the atomic weights of the elements, one can easily determine the formula weight. The number of gram formula weights (number of moles) can then be determined from the masses.

Exercise 3.2

Use conversion factors to solve the following:
1. Calculate the weight in grams of one mole of CO_2.
2. Calculate the number of moles of CO_2 in 56.0g of CO_2.
3. Calculate the weight in grams of 2.00×10^{22} molecules of CO_2.
4. Calculate the number of molecules in 22.0 grams of CO_2.

Exercise 3.2 Answers
1. One mole of CO_2 contains 1 mole of C atoms and two moles of oxygen atoms. The atomic weight of C is 12.01. The atomic weight of O is 16.00.

   1 mole C $\times \dfrac{12.01\text{g}}{\text{mole}} = 12.01$ g

   2 moles O $\times \dfrac{16.00\text{g}}{\text{mole}} = 32.00$ g

$$12.01g + 32.00g = 44.01g$$

The weight of one mole of CO_2 is 44.01g

2. $56.0g \times \dfrac{1 \text{ mole}}{44.01g} = 1.27$ mole

3. 2.00×10^{22} molecules $\times \dfrac{1 \text{ mole}}{6.02 \times 10^{23} \text{ molecules}} \times \dfrac{44.01g}{\text{mole}} = 1.46g$

4. $22.0g \times \dfrac{1 \text{ mole}}{44.01g} \times 6.02 \times 10^{23} \dfrac{\text{molecules}}{\text{mole}} = 3.01 \times 10^{23}$ molecules

2. Calculations Based on Formulas

Some types of calculations based on formulas were illustrated in the discussion of the mole. A determination of the weight percentages of each kind of atom in a compound is another type of calculation based on a chemical formula. The following is an example of such a calculation.

Calculate the percentages by weight of hydrogen and oxygen in water (H_2O).

The formula indicates that one mole of H_2O contains two moles of hydrogen atoms and one mole of oxygen atoms. The weight of one mole of water is 18.0g. The weight of hydrogen in one mole of water is:

2 moles H $\times 1.01 \dfrac{g}{\text{mole}} = 2.02g$ H

The percentage of hydrogen is: $\dfrac{2.02 \text{ g} \times 100}{18.0g} = 11.2\%$ H.

The weight of oxygen in one mole of water is:

1 mole O $\times \dfrac{16.0g}{\text{mole}} = 16.0g$

The percentage of oxygen is:

$\dfrac{16.0g}{18.0g} \times 100 = 88.8\%$

The use of analytical data and the percentage composition by weight to determine an empirical formula is illustrated in Examples 3.1, 3.2, and 3.3 in the text. The determination of a molecular formula is illustrated in Example 3.4 in the text.

Exercise 3.3
1. A 6.74g sample of an oxide of lead decomposed on heating at a high temperature to yield 6.11g of lead, the oxygen being set free. Calculate the percentage composition by weight and determine the empirical formula of the oxide of lead.
2. A 2.765g sample of a compound containing only C and H was burned completely in oxygen. The only products formed were 3.550g H_2O and 8.690g CO_2. From another experiment the molecular weight was found to be 70.0. Determine the molecular formula of the compound.

Exercise 3.3   Answers
1. The weight of oxygen in the oxide is the difference between the weight of the lead oxide and the lead that is set free.

   6.74g − 6.11g = 0.63g oxygen

   The percentage of lead in the lead oxide is $\frac{6.11g}{6.74g} \times 100 = 90.7\%$.

   The percentage of oxygen in the lead oxide is $\frac{0.63g}{6.74g} \times 100 = 9.3\%$.

   The percentage of oxygen could also be obtained by subtracting 90.7% from 100%.

   To determine the empirical formula, assume that we have 100g of the lead oxide. Therefore, we have 90.7g Pb and 9.3g O in 100g of the oxide. Calculate the number of moles of Pb atoms and O atoms as follows:

   $90.7g \times \frac{1 \text{ mole Pb}}{207.2g} = 0.438$ mole Pb

   $9.3g \times \frac{1 \text{ mole O}}{16.0g} = 0.581$ mole O

   To determine the mole relationship, divide the number of moles of O by the number of moles of Pb. This will give ratio of the number of moles of O per one mole of Pb.

   $\frac{0.581 \text{ mole O}}{0.438 \text{ mole Pb}} = 1.33 \frac{\text{mole O}}{\text{mole Pb}}$ or $PbO_{1.33}$

   To convert this ratio to integers, multiply by 3 to give the formula Pb_3O_4.

2. One method of solving this problem is to calculate the number of moles of C and H that are required to produce 8.690g CO_2 and 3.550g H_2O.

The number of moles of C in the original sample is:

$$8.690g \times \frac{1 \text{ mole } CO_2}{44.01g} \times \frac{1 \text{ mole C}}{1 \text{ mole } CO_2} = 0.1974 \text{ moles C}$$

The number of moles of H in the original sample is:

$$3.550g \times \frac{1 \text{ moles } H_2O}{18.00g} \times \frac{2 \text{ mole H}}{1 \text{ mole } H_2O} = 0.3944 \text{ moles H}$$

The mole ratio is $\frac{0.3944 \text{ moles H}}{0.1974 \text{ mole C}} = 1.998 \frac{\text{moles H}}{\text{mole C}}$ or 2 moles H per mole C. The empirical formula is CH_2 with a formula weight of $12 + 2 = 14$. Since the molecular weight is 70, $\frac{70}{14} = 5$, the molecular formula is C_5H_{10}. Note that this is different than writing $5CH_2$. The $5CH_2$ implies 5 moles with a formula unit of CH_2, not 1 mole with a formula of C_5H_{10}.

3. Writing Equations

A chemical equation is a way of describing a chemical reaction. To correctly represent a reaction, the following conditions must be met:
1. The formulas written on the left and on the right of the equation must correctly represent the reactants and products, respectively.
2. The coefficients in the equation must be consistent with the conservation of mass and conservation of charge.

The process of balancing equations is discussed and illustrated in Section 3.5 in the text.

Exercise 3.4

Balance the following equations.
1. $As^{3+}(aq) + S^{2-}(aq) \longrightarrow As_2S_3(s)$
2. $Ca(s) + H_2O(l) \longrightarrow Ca(OH)_2(s) + H_2(g)$
3. $Si(s) + Cl_2(g) \longrightarrow SiCl_4(l)$
4. $C_9H_8O(s) + O_2(g) \longrightarrow CO_2(g) + H_2O(g)$

Exercise 3.4 Answers
1. Two As atoms on the right require that there be two on the left. Three S atoms on the right require that there be three on the left. The balanced equation is

$$2As^{3+}(aq) + 3S^{2-}(aq) \longrightarrow As_2S_3(s)$$

Check to determine if mass is conserved and also if the net charge is the same on both sides. Note that on the left $2(+3) + 3(-2) = +6 - 6 = 0$. This agrees with the zero net charge on the right.

2. $Ca(s) + 2H_2O(l) \rightarrow Ca(OH)_2(s) + H_2(g)$
3. $Si(s) + 2Cl_2(g) \rightarrow SiCl_4(l)$
4. Recognize that one mole of C_9H_8O will produce nine moles of CO_2 and four moles of H_2O. The partially balanced equation is $C_9H_8O + O_2 \rightarrow 9CO_2 + 4H_2O$. The nine moles of CO_2 and four moles of H_2O will require 22 moles of oxygen atoms on the left. One mole of O atoms is present in the C_9H_8O leaving 21 moles to be furnished by the O_2, thus 10½ moles of O_2 are required. The balanced equation is:

$$C_9H_8O(s) + 10½ O_2(g) \rightarrow 9CO_2(g) + 4H_2O(g)$$

In many cases, the preferred form contains whole number coefficients. This can be obtained by multiplying all coefficients by 2.

$$2 C_9H_8O(s) + 21 O_2(g) \rightarrow 18 CO_2(g) + 8 H_2O(g)$$

Check for conservation of mass.

4. Calculations Based on Equations

The mole method of solving problems based on chemical equations is illustrated in Example 3.7 in the text. Remember that the coefficients in a balanced equation indicate the *relative* numbers of moles of reactants and products. The quantities of reactants available for reaction are not always in the same ratio as the quantities that actually react. The quantity of one of the reactants may be in excess; therefore, the yield of products will be dependent upon the "limiting" reagent. A direct comparison of the number of grams of reactants available will not necessarily lead to the correct conclusion as to which substance is in excess. An evaluation must be made in terms of the number of moles available and the mole relationships involved, as indicated by the balanced equation.

The actual yield of product is usually less than the theoretical yield that is calculated based on the chemical equation. A problem in which a theoretical yield and a percentage yield are calculated when one of the reactants is present in excess is illustrated in Example 3.8. These principles are also applied in the following exercise.

Exercise 3.5
1. Ammonia (NH_3) burns to produce nitrogen gas (N_2) and water vapor. Calculate the number of grams of nitrogen that can be

produced from (a) 42.0g of NH_3 with excess O_2 and (b) 42.0g NH_3 and 58.0g O_2. The unbalanced equation is:

$$NH_3(g) + O_2(g) \longrightarrow N_2(g) + H_2O(g)$$

2. Alcohol, C_2H_5OH, can be produced from glucose according to the equation:

$$C_6H_{12}O_6(aq) \longrightarrow 2C_2H_5OH(aq) + 2CO_2(g)$$

If the percentage yield of alcohol is 76.0%, how many grams of glucose are required to produce 1.00×10^2 g of alcohol?

3. A 3.00g sample of a crude iron sulfide ore, in which all the sulfur was present as FeS_2, was analyzed for iron as follows: The ore sample was treated with concentrated HNO_3 until all of the sulfur was converted to sulfuric acid, H_2SO_4. The H_2SO_4 was then reacted with $BaCl_2$ until all of the sulfate from the sulfuric acid was precipitated as $BaSO_4$. The $BaSO_4$ that was produced weighed 2.47g. Calculate the percentage of FeS_2 in the crude ore.

Exercise 3.5 Answers
1. a. First, balance the equation:

$$4NH_3(g) + 3\,O_2(g) \longrightarrow 2\,N_2(g) + 6\,H_2O(g)$$

Now, convert the amount of NH_3 given, 42.0g, first to moles NH_3, then to moles N_2 and finally to grams N_2.

$$42.0g\ NH_3 \times \frac{1\ \text{mole}\ NH_3}{17.0g\ NH_3} \times \frac{2\ \text{moles}\ N_2}{4\ \text{moles}\ NH_3} \times \frac{28.0g\ N_2}{1\ \text{mole}\ N_2} = 34.6g\ N_2.$$

b. First calculate the number of moles of each reactant available:

number of moles NH_3 present: $42.0g \times \dfrac{1\ \text{mole}\ NH_3}{17.0g} = 2.47$ mole NH_3

number of moles O_2 present: $58.0g \times \dfrac{1\ \text{mole}\ O_2}{32.0g} = 1.81$ mole O_2

Based on the balanced equation, calculate the number of moles of O_2 required to react with 2.47 moles NH_3.

$$2.47\ \text{mole}\ NH_3 \times \frac{3\ \text{mole}\ O_2}{4\ \text{mole}\ NH_3} = 1.85\ \text{mole}\ O_2\ \text{required}$$

Since only 1.81 mole of O_2 is present, not all of the NH_3 will react. The NH_3 is in excess and the calculation must be based on the "limiting" reagent, O_2 in this case. The final step in the calculation is to convert the number of moles of the limiting reagent to grams of product.

$$1.81 \text{ mole } O_2 \times \frac{2 \text{ mole } N_2}{3 \text{ mole } O_2} \times \frac{28.0 \text{g}}{\text{mole } N_2} = 33.8 \text{g } N_2$$

2. One approach to this problem is to calculate the number of grams of glucose required to produce 100.0g of alcohol if there is a 100% yield.

$$1.00 \times 10^2 \text{g } C_2H_5OH \times \frac{1 \text{ mole } C_2H_5OH}{46.0\text{g}} \times \frac{1 \text{ mole } C_6H_{12}O_6}{2 \text{ mole } C_2H_5OH} \times \frac{180.0\text{g}}{1 \text{ mole } C_6H_{12}O_6} = 1.96 \times 10^2 \text{g}.$$

A common error at this point is to multiply the percentage yield by the theoretical yield, $0.760 \times 196\text{g} = 149\text{g}$ and report 149g as the quantity of glucose required. If we think a moment, we know that if we start with 196g and the percentage yield is 76.0%, we can not produce 100 grams of alcohol by this reaction. We must start with more than 196 grams of glucose to produce 100 grams of alcohol if the yield is 76.0% of the theoretical. The question that must be answered is: 76.0% of what number is equal to 196g, or $0.760x = 196\text{g}$, $x = \frac{196 \text{ g}}{0.760} = 258\text{g}$. We must start with 258g of glucose to produce 100 grams of alcohol when the yield is 76.0%.

3. In a problem in which there is a series of reactions, it is not always necessary to write balanced equations for each step in the sequence. Since all of the sulfur in the original reactant appears in one final product, this problem can be solved based on the moles of sulfur as indicated in the chemical formulas that are given. We know that one mole of FeS_2 must produce two moles of $BaSO_4$ as long as all the sulfur in FeS_2 eventually takes the form of sulfate in $BaSO_4$. It makes no difference how many intermediate reactions are required to produce the sulfate from the FeS_2. Using conversion factors,

$$2.47\text{g } BaSO_4 \times \frac{1 \text{ mole } BaSO_4}{233\text{g}} \times \frac{1 \text{ mole } FeS_2}{2 \text{ mole } BaSO_4} \times \frac{120\text{g}}{\text{mole } FeS_2} = 0.636\text{g } FeS_2$$

Thus, 0.636g of FeS_2 is required to produce 2.47g of $BaSO_4$. The percentage of FeS_2 in the crude ore sample is:

$$\frac{0.636\text{g}}{3.00\text{g}} \times 100 = 21.2\%$$

SUGGESTED ASSIGNMENT III

Text: Chapter 3

Problems: Chapter 3, Numbers 3.1–3.7, 3.30, 3.34 3.36, 3.43

Solutions to Assigned Problems Set III

3.1

We need to determine the relative numbers of C and H atoms. We can calculate the number of GAWs of C and H or the number of moles of C and H atoms in one hundred grams of the compound. Based on the analysis, 100 grams of ethylene contains 85.6g C and 14.4g H.

$$85.6\text{g C} \times \frac{1 \text{ mole C}}{12.0\text{g C}} = 7.13 \text{ moles C}$$

$$14.4\text{g H} \times \frac{1 \text{ mole H}}{1.01\text{g H}} = 14.3 \text{ moles H}$$

Since the relative numbers of moles of C and H are exactly the same as the relative numbers of atoms of C and H, the atom ratio of H to C is

$$\frac{14.3 \text{ atoms H}}{7.13 \text{ atoms C}} = 2 \text{ atoms H/atom C}$$

The simplest formula is CH_2

3.2

The formula weight of the simplest formula of ethylene, CH_2, is

$12.0 + 2(1.01) = 14.02$

The approximate molecular weight, 30, is about twice the formula weight. The molecular formula of ethylene is C_2H_4.

3.3

a. The weight of one mole of CaO is 40.0g + 16.0g = 56.0g. Using 56.0g CaO/mole CaO as a conversion factor in the form 1 mole CaO/56.0g CaO,

$$19.6\text{g CaO} \times \frac{1 \text{ mole CaO}}{56.0\text{g CaO}} = 0.350 \text{ mole CaO}$$

b. Using 56.0g CaO/mole CaO as a conversion factor,

$$2.19 \text{ moles CaO} \times \frac{56.0\text{g CaO}}{\text{mole CaO}} = 123\text{g CaO}$$

3.4

The weights of C, H, and O in one mole of aspirin are:

$$9 \text{ moles C} \times \frac{12.0\text{g C}}{\text{mole C}} = 108 \text{ g C}$$

$$8 \text{ moles H} \times \frac{1.0\text{g H}}{\text{mole H}} = 8.0 \text{ g H}$$

$$4 \text{ moles O} \times \frac{16.0\text{g O}}{\text{mole O}} = 64.0 \text{ g O}$$

108g + 8.0g + 64.0g = 180 g = the weight of 1 mole $C_9H_8O_4$

$$\% \text{ C} = \frac{108\text{g}}{180\text{g}} \times 100 = 60.0\%$$

$$\% \text{ H} = \frac{8.0\text{g}}{180\text{g}} \times 100 = 4.4\%$$

$$\% \text{ O} = \frac{64.0\text{g}}{180\text{g}} \times 100 = 35.6\%$$

3.5

When a substance burns in air, it reacts with oxygen, O_2. Write the correct formulas for the reactants and products.

$$PH_3 + O_2 \longrightarrow H_2O + P_4O_{10} \text{ (unbalanced)}$$

Four moles of PH_3 are required to produce 1 mole of P_4O_{10}. Four moles of PH_3 will contain enough hydrogen to produce six moles of H_2O.

$$4PH_3 + O_2 \longrightarrow 6H_2O + P_4O_{10} \text{ (unbalanced)}$$

Sixteen moles of oxygen atoms or eight moles of O_2 are required. The proper notations for gases, pure liquids, and solids are also indicated in the balanced equation.

$$4PH_3(g) + 8O_2(g) \rightarrow 6H_2O(l) + P_4O_{10}(s)$$

3.6

The coefficients in the balanced equation represent the relative numbers of moles of reactants and products. We can use the coefficients in the equation, along with the number of grams per mole, as conversion factors to solve these problems.

a. $1.16 \text{ moles } P_4H_{10} \times \dfrac{4 \text{ moles } PH_3}{1 \text{ mole } P_4H_{10}} = 4.64 \text{ moles } PH_3$

b. $0.198 \text{ mole } O_2 \times \dfrac{6 \text{ moles } H_2O}{8 \text{ moles } O_2} \times \dfrac{18.0 \text{g } H_2O}{\text{mole } H_2O} = 2.67 \text{g } H_2O$

c. $12.0 \text{g } PH_3 \times \dfrac{1 \text{ mole } PH_3}{34.0 \text{g } PH_3} \times \dfrac{8 \text{ moles } O_2}{4 \text{ moles } PH_3} \times \dfrac{32.0 \text{g } O_2}{1 \text{ mole } O_2} = 22.6 \text{g } O_2$

3.7

a. First convert the number of grams to the number of moles of each reactant.

$6.80 \text{g } PH_3 \times \dfrac{1 \text{ mole } PH_3}{34.0 \text{g } PH_3} = 0.200 \text{ mole } PH_3 \text{ available}$

$6.40 \text{g } O_2 \times \dfrac{1 \text{ mole } O_2}{32.0 \text{g } O_2} = 0.200 \text{ mole } O_2 \text{ available}$

Referring to the balanced equation in Problem 3.5, we see that two moles of oxygen are required for every mole of PH_3 that reacts. The limiting reagent is O_2.

b. The number of grams of P_4O_{10} that can be produced is based on the amount of O_2, the limiting reagent. Using conversion factors,

$0.200 \text{ mole } O_2 \times \dfrac{1 \text{ mole } P_4O_{10}}{8 \text{ moles } O_2} \times \dfrac{284 \text{g } P_4O_{10}}{1 \text{ mole } P_4O_{10}} = 7.10 \text{g } P_4O_{10}$

c. $\dfrac{5.20 \text{g}}{7.10 \text{g}} \times 100 = 73.2\%$

3.30

The weight of C in 1.642mg lindane can be calculated from the weight of CO_2 produced.

$$1.491\text{mg } CO_2 \times \frac{12.0\text{g C}}{44.0\text{g } CO_2} = 0.407\text{mg C}$$

By a similar calculation, the weight of H in the sample can be determined.

$$0.305\text{mg } H_2O \times \frac{2.02\text{g H}}{18.0\text{g } H_2O} = 0.0342\text{mg H}$$

The weight of Cl can be determined by taking the difference.

$$1.642\text{mg} - (0.407 + 0.0342)\text{mg} = 1.201\text{mg Cl}$$

The relative number of moles of each atom can be calculated from the masses of C, H, and Cl.

$$4.07 \times 10^{-4} \text{g C} \times \frac{1 \text{ mole C}}{12.0\text{g C}} = 3.39 \times 10^{-5} \text{ mole C}$$

In a similar fashion, convert to moles of H and Cl.

$$3.42 \times 10^{-5} \text{g H} \times \frac{1 \text{ mole H}}{1.01\text{g H}} = 3.39 \times 10^{-5} \text{ mole H}$$

$$1.20 \times 10^{-3} \text{g Cl} \times \frac{1 \text{ mole Cl}}{35.45\text{g Cl}} = 3.39 \times 10^{-5} \text{ mole Cl}$$

The simplest formula is CHCl.

3.34

The weight of one mole of $(NH_4)_3PO_4$ is determined as follows:

$$3 \text{ moles N} \times \frac{14.0\text{g N}}{1 \text{ mole N}} = 42.0\text{g}$$

$$12 \text{ moles H} \times \frac{1.01\text{g H}}{1 \text{ mole H}} = 12.1\text{g}$$

$$1 \text{ mole P} \times \frac{31.0\text{g P}}{1 \text{ mole P}} = 31.0\text{g}$$

4 moles O × 16.0g O = 64.0g

42.0g + 12.1g + 31.0g + 64.0g = 149.1g

a. $1.00g \times \dfrac{1 \text{ mole}}{149g} = 6.71 \times 10^{-3}$ mole

b. $0.204 \text{ mole} \times \dfrac{149g}{1 \text{ mole}} = 30.4g$

c. The number of N atoms in 6.71×10^{-3} mole $(NH_4)_3PO_4$ is:

6.71×10^{-3} mole $(NH_4)_3PO_4 \times \dfrac{3 \text{ moles N atoms}}{1 \text{ mole } (NH_4)_3PO_4} \times$

$\dfrac{6.02 \times 10^{23} \text{ atoms}}{\text{mole N atoms}} = 1.21 \times 10^{22}$ atoms

The number of N atoms in 0.204 mole $(NH_4)_3PO_4$ is

0.204 mole $(NH_4)_3PO_4 \times \dfrac{3 \text{ moles N atoms}}{1 \text{ mole } (NH_4)_3PO_4} \times$

$\dfrac{6.02 \times 10^{23} \text{ atoms}}{\text{mole N atoms}} = 3.68 \times 10^{23}$ atoms

3.36

a. $B_2H_6(l) + 3O_2(g) \longrightarrow B_2O_3(s) + 3H_2O(l)$

b. $2Ag^+(aq) + CO_3{}^{2-}(aq) \longrightarrow Ag_2CO_3(s)$

c. $C_3H_5N_3O_9(l) \longrightarrow 3CO_2(g) + 1\frac{1}{2}N_2(g) + 2\frac{1}{2}H_2O(l) + \frac{1}{4}O_2(g)$

$4C_3H_5N_3O_9(l) \longrightarrow 12CO_2(g) + 6N_2(g) + 10H_2O(l) + O_2(g)$

3.43

For a 100% yield, with no excess KSCN, he would need:

$20g \times \dfrac{1 \text{ mole } [Co(NH_3)_5SCN]Cl_2}{273g} \times \dfrac{1 \text{ mole } [Co(NH_3)_5Cl]Cl_2}{1 \text{ mole } [Co(NH_3)_5SCN]Cl_2} \times$

$\dfrac{250g}{\text{mole}} = 18g [Co(NH_3)_5Cl]Cl_2$

$20g \times \dfrac{1 \text{ mole } [Co(NH_3)_5SCN]Cl_2}{273g} \times \dfrac{1 \text{ mole KSCN}}{1 \text{ mole } [Co(NH_3)_5SCN]Cl_2} \times$

$\dfrac{97g}{\text{mole KSCN}} = 7.1g$

For a 55% yield:

$18g/0.55 = 33g\ [Co(NH_3)_5Cl]Cl_2$
$7.1g/0.55 = 13g\ KSCN$

With 60% excess KSCN:

$0.60 \times 13g = 7.8g;\ 13\ g + 7.8g = 21g\ KSCN$

UNIT 4

Thermochemistry

The goal of this unit is to develop an understanding of the quantitative relationships associated with energy changes in chemical reactions. In addition, principles of thermochemistry will be applied to problems associated with energy sources and uses.

INSTRUCTIONAL OBJECTIVES

4.1 The Enthalpy Change: ΔH
  1. Distinguish between exothermic and endothermic changes.
  2. Review the definition of calorie and kilocalorie; relate these to other units of energy. (Chapter 1 in the text, Table 1.2)
  3. Define enthalpy, H, and change in enthalpy, ΔH.
  *4. Recognize that for a constant pressure process, Q_p = H products − H reactants = ΔH reaction.
  5. Recognize that $\Delta H < 0$ for an exothermic process; $\Delta H > 0$ for an endothermic process.
  6. Recognize that the values of state properties, such as H, V, P, and T, are independent of their history.
  7. Realize that the enthalpy of a substance is directly proportional to the mass of the substance.

4.2 Thermochemical Equations
  *1. Interpret the conventions used in writing thermochemical equations.
  2. Recognize that the value of ΔH is dependent upon the states of the reactants and products.
  *3. Using thermochemical equations, calculate ΔH for a reaction when given the amounts of products or reactants. (Prob. 4.1a)
  *4. Using thermochemical equations, calculate the amount of reactant or product that will be required to obtain a given value of ΔH. (Prob. 4.1b)

5. Realize that ΔH for a reaction is equal in magnitude but opposite in sign, to ΔH for the reverse reaction.
*6. Apply Hess's Law to calculate ΔH of a reaction when given enthalpy changes for other reactions. (Probs. 4.2, 4.3, 4.27)

4.3 Heats of Formation
1. Define molar heat of formation, ΔH_f.
2. Realize that the heat of formation of any elementary substance is zero.
*3. Calculate ΔH of a reaction from heats of formation; calculate heats of formation from heats of reaction. (Probs. 4.2, 4.27, 4.29)

4.4 Bond Energies (Enthalpies)
1. Define bond energy, ϵ.
*2. Calculate ΔH for a reaction when given the values of bond energies. (Prob. 4.3)
3. Realize that bond energy values may vary slightly, depending on the species in which the bond is found.

4.5 Measurement of Heat Flow—Calorimetry
1. Describe the principles involved in measuring heat flow with a simple calorimeter and with a bomb calorimeter.
2. Explain what is meant by heat capacity, C, of a calorimeter.
*3. Calculate the fourth when given any three of the following:
   A. heat flow (probs. 4.4a, 4.5a)
   B. amount of substance in grams or in moles
   C. temperature change, or initial or final temperature
   D. specific heat of a substance or heat capacity of a calorimeter (Prob. 4.4b)

4.6 The First Law of Thermodynamics
1. Distinguish between system and surroundings.
*2. Apply the proper sign conventions to indicate the direction of energy flow in the form of heat or work. (Prob. 4.35)
3. State, both in words and as an equation, the First Law of Thermodynamics.
4. Realize that change in internal energy, ΔE, is a state property, and that Q and W are not state properties.
*5. Relate change in internal energy, ΔE, to heat flow, Q, and work, W. (Prob. 4.6)
*6. Recognize that W = PΔV for processes occuring at constant pressure conditions and that PΔV = $(0.0020)(\Delta n_g)$ (T) for a reaction involving gases.
*7. Use Equation 4.24 to relate ΔH and ΔE. (Prob. 4.7)

4.7 Sources of Energy
1. List some practices that may be employed to conserve our supply of fossil fuels.

2. Describe some possible alternatives to the use of fossil fuels and the problems associated with these alternatives.

SUMMARY

1. Enthalpies, Enthalpy Changes, and Thermochemical Equations

An equation that specifies the heat flow associated with a chemical reaction is called a thermochemical equation. For constant pressure processes, the heat flow, Q_p, is equal to the difference between the enthalpy (heat content) of the products and that of the reactants. Thermochemical equations are interpreted in terms of the number of moles of a substance. The enthalpy of a substance is also related to its physical state. Although the enthalpy of a substance is somewhat temperature dependent, the temperature effect on the enthalpies of individual substances tends to cancel out when measuring ΔH for a reaction.

Hess's Law is based on the fact that enthalpy is a state property; therefore, ΔH is independent of the path followed from an initial to a final state. This law is applied in the calculation of ΔH for an overall reaction by adding ΔH values for a series of related reactions. Hess's Law also relates ΔH of a reaction to the heats of formation (ΔH_f) of the products and reactants. The sign and magnitude of ΔH is related to bond energies of the products and reactants. Again, Hess's Law may be used to calculate the ΔH for a reaction from average bond energy values.

A calorimeter is a device used to measure the heat flow in a process. When the contents of the calorimeter are kept at a constant pressure, the heat flow is equal to ΔH for the process. The heat flow, Q, for this type of calorimeter is related to the specific heat, S. H., mass, and temperature change, Δt, by the expression $Q = (S. H.) (m) (\Delta t)$.

In a bomb calorimeter, heat flow is measured at conditions of constant volume. The heat evolved in a reaction is equal to the heat absorbed by the calorimeter plus the heat absorbed by the water. The heat capacity of the calorimeter must be determined in a separate experiment. For an exothermic reaction in a constant pressure calorimeter, the heat released in the reaction is equal to the amount of heat absorbed by the water.

Exercise 4.1

Octane, C_8H_{18}, is one of the major constitutents in gasoline. The value of ΔH for the combustion of octane is -1300 kcal/mole. The equation for the reaction is

$$C_8H_{18}(l) + 12\tfrac{1}{2} O_2(g) \longrightarrow 8CO_2(g) + 9 H_2O(l); \Delta H = -1300 \text{ kcal}$$

1. Calculate the amount of heat evolved when one gallon (about 2860g) of octane burns.
2. How many grams of octane will have to be burned to produce one kcal of heat?
3. Using Table 4.1 and the information above, calculate the heat of formation of C_8H_{18}(l).
4. Use Table 4.2 to estimate ΔH for the following reaction (all substances are gases):

$$2 \; H-\underset{\underset{H}{|}}{\overset{\overset{H}{|}}{C}}-\underset{\underset{H}{|}}{\overset{\overset{H}{|}}{C}}-\underset{\underset{H}{|}}{\overset{\overset{H}{|}}{C}}-\underset{\underset{H}{|}}{\overset{\overset{H}{|}}{C}}-H \rightarrow H-\underset{\underset{H}{|}}{\overset{\overset{H}{|}}{C}}-\underset{\underset{H}{|}}{\overset{\overset{H}{|}}{C}}-\underset{\underset{H}{|}}{\overset{\overset{H}{|}}{C}}-\underset{\underset{H}{|}}{\overset{\overset{H}{|}}{C}}-\underset{\underset{H}{|}}{\overset{\overset{H}{|}}{C}}-\underset{\underset{H}{|}}{\overset{\overset{H}{|}}{C}}-\underset{\underset{H}{|}}{\overset{\overset{H}{|}}{C}}-\underset{\underset{H}{|}}{\overset{\overset{H}{|}}{C}}-H + H_2$$

5. A 1.000g sample of octane is burned in a bomb calorimeter containing 1525g of water. The temperature rises from 21.42 to 27.66°C. If the heat of combustion of octane is 1300 kcal/mole, calculate the heat capacity of the bomb calorimeter in cal/°C.

Exercise 4.1 Answers

1. One mole of octane releases 1300 kcal. The weight of one mole of octane is 114g. Using conversion factors,

$$2860g \times \frac{1 \; mole}{114g} \times \frac{-1300 \; kcal}{1 \; mole} = -3.26 \times 10^4 \; kcal$$

When one gallon of octane burns, 3.26×10^4 kcal of heat is released.

2. Use conversion factors as follows:

$$1.00 \; kcal \times \frac{1 \; mole}{1300 \; kcal} \times \frac{114g}{1 \; mole} = 8.77 \times 10^{-2} g$$

3. The equation for burning octane is

$$C_8H_{18}(l) + 12\tfrac{1}{2} O_2(g) \longrightarrow 8CO_2(g) + 9H_2O(l); \; \Delta H = -1300 \; kcal$$

Since we know ΔH for the reaction, we can solve for ΔH_f reactants in the following expression. Remember, ΔH_f of an elementary substance is zero.

$$\Delta H = \Sigma \Delta H_f \; products - \Sigma \Delta H_f \; reactants$$

$$\Delta H = 8\Delta H_f CO_2(g) + 9 \; \Delta H_f H_2O(l) - \Delta H_f C_8H_{18}(l)$$

$$-1300 \; kcal = 8(-94.1 \; kcal) + 9 \; (-68.3 \; kcal) - \Delta H_f C_8 H_{18}(l)$$

$\Delta H_f C_8 H_{18}(l) = -752.8$ kcal -614.7 kcal $+ 1300$ kcal $= -68$ kcal

4. One approach is to calculate the energy required to break the bonds in the reactant molecules, ΔH_1 and then calculate the energy released in the bond forming process, ΔH_2. The bond-breaking process for two moles of $C_4 H_{10}$ requires:

$\Delta H_1 = 20\epsilon C - H + 6 \epsilon C - C = 20(99 \text{ kcal}) + 6(83 \text{ kcal}) = 2478$ kcal

The bond-forming process releases:

$\Delta H_2 = 7\epsilon C - C + 18 \epsilon C - H + \epsilon H - H = 7(-83 \text{ kcal}) + 18 (-99$ kcal$) + (-104$ kcal$) = -2467$ kcal

Using Hess's Law: $\Delta H = \Delta H_1 + \Delta H_2 = 2478$ kcal $- 2467$ kcal $= 11$ kcal

5. First calculate the amount of energy released when one gram of octane burns.

$1300 \frac{\text{kcal}}{\text{mole}} \times \frac{1 \text{ mole}}{114g} = 11.4$ kcal/g $= 11,400$ cal/g; $Q = -11,400$ cal

Using $Q = -(Q$ water $+ Q$ bomb$)$; and $\Delta t = 27.66°C - 21.42°C = 6.24°C$; $-11,400$ cal $= - \left(1.00 \frac{\text{cal}}{\text{g}°\text{C}} \times m_{H_2O} \times \Delta t + C \times \Delta t \right)$

$-11,400$ cal $= - \left(1.00 \frac{\text{cal}}{\text{g}°\text{C}} \times 1525g \times 6.24°C + C \times 6.24°C \right)$

$-11,400$ cal $= -9516$ cal $-(6.24°C)C$

$C(6.24°C) = 1900$ cal

$C = 300 \frac{\text{cal}}{°\text{C}}$

2. The First Law of Thermodynamics

Energy effects can be classified as heat (Q) and work (W). Just as heat flow out of or into a system may be indicated by sign conventions, work done by or on a system may be specified by sign conventions. The change in internal energy, ΔE, of a system is specified by the expression: $\Delta E_{system} = E_{final} - E_{initial} = Q - W$. Since ΔE depends only on the initial and final states of a system and not on its path, E is a state property. Since Q and W can each have any value, they are not state properties.

When applied to a reaction system, $\Delta E_{reaction} = E_{products} - E_{reactants}$. If the reaction occurs at constant volume conditions, no work associated with a volume change is involved, $W = 0$ and $\Delta E = Q$. For a reaction at constant pressure conditions, work is associated with a volume change and $W = P\Delta V$. The heat flow Q for constant pressure conditions is $Q = \Delta E + P\Delta V$. Since $\Delta H = Q$ at constant pressure, $\Delta H = \Delta E + P\Delta V$. The $P\Delta V$ term is quite small for reactions involving liquids and solids. However, for a reaction that produces a change in the number of moles of gas, the work term, $P\Delta V$, may be significant.

Exercise 4.2

The value of ΔH is -366 kcal/mole of NH_3 for the following reaction at 25°C.

$$4NH_3(g) + 3O_2(g) \rightarrow 2N_2(g) + 6H_2O(l)$$

1. What is the value of ΔN_g?
2. Calculate ΔE per mole of $NH_3(g)$ for the reaction.
3. Determine the heat flow, Q_p, for the reaction when one mole of NH_3 is burned in an open flame at 25°C.
4. Determine the heat flow, Q_v, for the reaction when one mole of NH_3 is burned in a bomb calorimeter.
5. Explain why the two values of Q calculated in 3 and 4 are different.

Exercise 4.2 Answers
1. There are two moles of gaseous product, seven moles of gaseous reactants, $\Delta N_g = 2 - 7 = -5$
2. $\Delta E = \Delta H - P\Delta V$; $P\Delta V = (0.0020)(T)(\Delta n_g) = (0.0020)(298)(-5) = -2.98$ kcal; $\Delta E = -366$ kcal $- (-3$ kcal$) = -366$ kcal $+ 3$ kcal $= -363$ kcal
3. $Q_p = \Delta H = -366$ kcal
4. $Q_v = \Delta E = -363$ kcal
5. No work of expansion or contraction is done in the constant volume of a bomb calorimeter, $Q_v = \Delta E$. When the NH_3 is burned in an open flame, there is a contraction. Three kcal of work are done by the surroundings on the system.

3. Sources of Energy

The dwindling supply of petroleum and natural gas reserves is a major problem facing the United States. Several suggestions have been made to conserve these reserves. Over the long term, however, it appears that solutions to the enegy problem will require that the country develop some alternatives to energy derived from fossil fuels. The development of such alternatives will require that we increase our efforts in research, engineering, and technology to develop practical, economical sources that are safe for man and his environment.

SUGGESTED ASSIGNMENT IV

Text: Chapter 4

Problems: Numbers 4.1–4.7, 4.27, 4.29, 4.35, 4.37

Solutions to Assigned Problems Set IV

4.1

a. From the equation, ΔH for the formation of two moles of MgO is -287.6 kcal. Using conversion factors:

$$1.00\text{g MgO} \times \frac{1 \text{ mole MgO}}{40.3 \text{ MgO}} \times \frac{-287.6 \text{ kcal}}{2 \text{ moles MgO}} = -3.57 \text{ kcal}$$

b. $1 \text{ kcal} \times \dfrac{1\text{g}}{3.57 \text{ kcal}} = 0.280\text{g}$

4.2

$\Delta H = \Sigma \Delta H_f \text{ products} - \Sigma \Delta H_f \text{ reactants}$

$= 4\Delta H_f CO_2(g) + 2\Delta H_f H_2O(l) - 2\Delta H_f C_2H_2(g)$

$= 4(-94.1 \text{ kcal}) + 2(-68.3 \text{ kcal}) - 2(+54.2 \text{ kcal}) = -621 \text{ kcal}$

4.3

ΔH_1 for the bond-breaking is:

$\Delta H_1 = 3(\epsilon N-H) + \epsilon Cl-Cl = 3(93 \text{ kcal}) + 58 = +337 \text{ kcal}$

ΔH_2 for bond formation is:

$\Delta H_2 = 2(\epsilon N-H) + \epsilon N-Cl + \epsilon H-Cl = 2(-93 \text{ kcal}) + (-48 \text{ kcal}) + (-103 \text{ kcal}) = -337 \text{ kcal}$

By Hess's Law: $\Delta H = H_1 + H_2 = +337 \text{ kcal} - 337 \text{ kcal} = 0$

4.4

a. The amount of heat absorbed by the water is equal to the amount of heat lost by the metal. The amount of heat gained by the water, Q, when $\Delta t = (24.0 - 20.0)°C = 4.0°C$ is related by Equation 4.15.

$$Q \text{ water} = (S.H.)(m)(\Delta t) = \left(1.00 \frac{cal}{g°C}\right)(30.0g)(4.0°C) = +120 \text{ cal}$$

b. Q metal is -120 cal. Applying Equation 4.15 to the metal,

$$Q \text{ metal} = (S.H.)(m)(\Delta t) = -120 \text{ cal} = (S.H.)(10.0g)(24°C - 90°C)$$

$$\text{S.H. Metal} = \frac{-120 \text{ cal}}{(10.0g)(-66°C)} = 0.18 \text{ cal/g°C}$$

4.5

a. Equation 4.16, $Q = -(Q_{H_2O} + Q_{bomb})$ relates the amount of heat released in the combustion to the amount of heat absorbed by the water plus that absorbed by the calorimeter.

$Q_{H_2O} = (S.H.)(m_{H_2O})(\Delta t)$; $Q_{bomb} = (C)(\Delta t)$. $\Delta t = 33.84°C - 25.00°C = 8.84°C$. Substituting,

$$Q = -\left(1.00 \frac{cal}{g°C} \times 1200g \times 8.84°C + 200 \frac{cal}{°C} \times 8.84°C\right) = -12.4 \text{ kcal}$$

The 12.4 kcal of heat evolved by the reaction is absorbed by the water and the bomb.

b. For one mole of ethane: $Q = -12.4 \frac{kcal}{g} \times 30.0 \frac{g}{mole} = -372 \frac{kcal}{mole}$

4.6

Since the system evolves heat, $Q = -1.62$ kcal. The system is doing work; therefore, $W = +0.36$ kcal.

$\Delta E = Q - W = -1.62 \text{ kcal} - 0.36 \text{ kcal} = -1.98 \text{ kcal}$

4.7

a. For a reaction at constant pressure conditions, $Q_p = \Delta H = -530.7$ kcal

b. In this reaction there are three moles of gaseous products and a total of six moles of gaseous reactants.

Δn_g = number of moles gaseous products - number of moles gaseous reactants = 3 - 6 = -3

c. Substituting into Equation 4.24:

$\Delta H = \Delta E + 0.60 \, \Delta n_g = -530.7$ kcal $= \Delta E + 0.60(-3)$;

$\Delta E = -528.9$ kcal

4.27

The method of solution to this problem is to arrange the given equations so that when these equations are added, the equation sought results. $Mn(s) + O_2(g) \rightarrow MnO_2(s)$; $H_f = ?$ This can be done by reversing the first equation, multiplying this entire equation and the ΔH value by two, and adding this to the second equation. Note that the sign of ΔH is changed when the first equation is written in the reverse direction,

$2(MnO(s) + \tfrac{1}{2}O_2(g) \rightarrow MnO_2(s))$; $\Delta H = 2(-32.5$ kcal$)$

$\underline{MnO_2(s) + Mn(s) \rightarrow 2MnO(s) \; ; \quad \Delta H = -59.5 \text{ kcal}}$
$Mn(s) + O_2(g) \rightarrow MnO_2(s) \; ; \quad \Delta H_f = -124.5$ kcal

4.29

For the reaction in Problem 4.26, $\Delta H = -8.00 \times 10^3$ kcal.

$\Delta H = \Sigma \Delta H_f$ products $- \Sigma \Delta H_f$ reactants

$\Delta H = 57 \Delta H_f CO_2(g) + 52 \Delta H_f H_2O(l) - \Delta H_f C_{57}H_{104}O_6(s) - 8.00 \times 10^3$ kcal $= 57(-94.1$ kcal$) + 52(-68.3$ kcal$) - \Delta H_f C_{57}H_{104}O_6(s)$

$\Delta H_f C_{57}H_{104}O_6(s) = -5364$ kcal $- 3552$ kcal $+ 8000$ kcal $= -9.2 \times 10^2$ kcal/mole

4.35

The relationship applied in this problem is $\Delta E = Q - W$.

a. Write the above equation in the form $W = Q - \Delta E$. When $\Delta E = -56$ cal, $W = 90$ cal $-(-56$ cal$) = +146$ cal. Since W is positive, the system is doing work on the surroundings.

b. $W = -27$ cal $- (-56$ cal$) = +29$ cal. Again, the system is doing work on the surroundings.

c. If 54 cal of heat is absorbed, $Q = +54$ cal.

$W = +54$ cal $- (-56$ cal$) = 110$ cal

4.37

a. In 1900: $\dfrac{0.26 \times 10^{16} \text{ kcal}}{7.6 \times 10^{7} \text{ persons}} = 3.4 \times 10^{7} \dfrac{\text{kcal}}{\text{person}}$

   In 1970: $\dfrac{1.8 \times 10^{16} \text{ kcal}}{20.4 \times 10^{7} \text{ persons}} = 8.8 \times 10^{7} \dfrac{\text{kcal}}{\text{person}}$

b. $(3.4 \times 10^{7}$ kcal/person$)(20.4 \times 10^{7}$ persons$) = 6.9 \times 10^{15}$ kcal

c. $(8.8 \times 10^{7}$ kcal/person$)(7.6 \times 10^{7}$ persons$) = 6.7 \times 10^{15}$ kcal

Note that this suggests that about one half of the increase in energy consumption is accounted for by the increase in population.

UNIT 5

The Physical Behavior of Gases

The goal of this unit is to become familiar with the laws governing the physical behavior of gases and to develop a model (theory) to explain the characteristics of gases.

INSTRUCTIONAL OBJECTIVES

5.1 General Properties of Gases
 1. List some general properties of gases.

5.2 Atmospheric Pressure
 1. Discuss the principles upon which a barometer operates.
 2. Describe a manometer and explain how it is used. (Prob. 5.25)

5.3 The Ideal Gas Law
 1. State the Ideal Gas Law and show how it implies:
   a. Boyle's Law.
   b. Charles and Gay-Lussac's Law.
   c. Avogadro's Law.
 *2. Evaluate the magnitude and the proper units for the gas constant R.

5.4 Using the Ideal Gas Law
 *1. Determine the effect of a change in conditions upon a particular variable (e.g., the effect of change in T and or P on volume).
 *2. Evaluate one variable (P, V, T, or n) when given the values of three of the four. (Probs. 5.1c, 5.28, 5.36a)
 *3. Determine the density of a particular gas as a function of T and P. (Prob. 5.1d, 5.30)

*4. Calculate the molecualr weight of gas from its density (or the mass of a given volume) at a known P and T. (Prob. 5.31)
*5. Apply the Law of Combining Volumes. (Prob. 5.33)

5.5 Mixtures of Gases: Dalton's Law of Partial Pressures
*1. State and apply Dalton's Law to obtain partial pressures of gases in a mixture. (Probs. 5.2, 5.36b)
*2. Recognize that this law is applied to calculations involving the collection of gases by water displacement. (Prob. 5.32)

5.6 Real Gases
1. Interpret deviations from the Ideal Gas Law in terms of molecular behavior.

5.7 Kinetic Theory of Gases
1. State the postulates of the kinetic theory of gases.
*2. Employ Graham's Law to determine the molecular weight of an unknown gas, or the relative rates of effusion when suitable data are given. (Prob. 5.40)
*3. Calculate average molecular speeds or energies of translation by use of Equations 5.13 and 5.14. (Probs. 5.3, 5.38)
4. Explain the significance of the distribution of molecular speeds (or energies) of a gas sample at two different temperatures. (Fig. 5.13)

SUMMARY

1. Gas Pressure Measurements

Pressure is a force per unit area. Gas pressures are often expressed in terms of the height of a mercury column that exerts the same pressure as the gas. A mercury column 1 mm high exerts a pressure of one mm Hg; 760 mmHg are equal to one standard atmosphere. A barometer is used to measure the atmospheric pressure. A manometer is usually used to measure the pressure of a gas in a closed container, the difference between mercury levels being directly proportional to the difference between the two gas pressures.

2. The Gas Laws

The four properties used to describe gas systems are pressure, temperature, volume, and number of moles. Boyle's Law states that for a fixed quantity of gas at a constant temperature, PV equals a constant. This relationship indicates that an increase in pressure on a gas will decrease the volume of the gas.

The volume of a mass of gas, at constant pressure, is directly proportional to the absolute temperature. This is a statement of Charles' Law.

The Ideal Gas Law, PV = nRT, combines the various gas laws into one expression. The value of the gas constant, R, common to all gases, is 0.0821 liter atm/mole°K. When using this value of R, the variables must be in units that are consistent with the units of R.

The extension of the gas laws to the behavior of gases in chemical reactions is summarized in Gay-Lussac's Law of Combining Volumes. At constant temperature and pressure, the volumes of gaseous reactants or products in a chemical reaction are in the ratio of small whole numbers. In fact, at conditions of constant temperature and pressure, the volume ratios are the same as the ratios of the coefficients in the balanced equation.

The gas laws have many applications as illustrated in the following exercise and the examples in Section 5.4 of the text.

Exercise 5.1
1. A sealed 125 ml flask contains nitrogen (N_2) gas at a pressure of 785 mmHg and a temperature of 35°C. Calculate the number of grams of nitrogen in the flask.
2. A 1.35g sample of an unknown gas occupies 475 ml. The pressure exerted by the gas in 748 mmHg at 96°C. Calculate the molecular weight of the gas.
3. When 45.0 liters of C_2H_6 gas are burned to gaseous CO_2 and H_2O at 777 mmHg and 153°C, how many liters of CO_2 gas are produced?

Exercise 5.1 Answers
1. Use the Ideal Gas Law to calculate the number of moles of N_2. Remember to change the temperature to °K, the pressure to atmospheres, and the volume to liters if you use R = 0.0821 liter atm/°K mole.

$$785 \text{ mmHg} \times \frac{1 \text{ atm}}{760 \text{ mmHg}} = 1.03 \text{ atm}; \quad 273 + 35° = 308°K;$$

125 ml = 0.125 liter

$$n = \frac{PV}{RT} = \frac{(1.03 \text{ atm})(0.125 \text{ liter})}{(0.0821 \text{ liter atm}/°K \text{ mole})(308°K)} = 5.09 \times 10^{-3} \text{ moles}$$

$$5.09 \times 10^{-3} \text{ moles} \times \frac{28.0g}{\text{mole}} = 1.43 \times 10^{-1} g$$

2. This problem may be solved as in Example 5.6 of the text or in two steps as illustrated here. First use the Ideal Gas Law to calculate the number of moles of gas.

$$n = \frac{PV}{RT} = \frac{\left(\frac{748}{760}\text{ atm}\right)(0.475 \text{ liter})}{(0.0821 \text{ liter atm/}°K \text{ mole})(369°K)} = 1.54 \times 10^{-2} \text{ moles}$$

Then calculate the number of grams per mole.

$$\frac{1.35 \text{g}}{1.54 \times 10^{-2} \text{ mole}} = 87.7 \text{ g/mole}$$

3. The balanced equation is

$$2C_2H_6(g) + 7O_2(g) \longrightarrow 4CO_2(g) + 6H_2O(g)$$

According to Gay-Lussac's Law, the volume ratios for gases measured at constant pressure and temperature are the same as the mole ratios in the balanced equation.

$$45.0 \text{ liters } C_2H_6 \times \frac{4 \text{ liters } CO_2}{2 \text{ liters } C_2H_6} = 90.0 \text{ liters } CO_2$$

Since the conditions of the reaction produce water vapor, the volume of gaseous water produced can be calculated in a similar way. If liquid water is produced or if all of the gases are not at the same temperature and pressure conditions, Gay-Lussac's Law cannot be applied.

In a mixture of non-reacting gases, each gas behaves independently of the other gases that are present. Therefore, each gas exerts a pressure (called the partial pressure) that it would exert if it alone occupied the volume. The total pressure in a container is equal to the sum of the partial pressures of the gases in the mixture. This is a statement of Dalton's Law. It follows that in such a mixture of gases, the partial pressure of a gas is directly proportional to the mole fraction of that gas.

$$\text{The mole fraction of gas A} = \frac{\text{number of moles A}}{\text{total number moles}}$$

An Application of Dalton's Law is the calculation of the pressure of a gas that is collected over a liquid. This application is illustrated in Example 5.9 in the text.

Graham's Law of Effusion relates the rate at which gaseous molecules move to the molecular masses of the molecules. The law is summarized in the form

$$\frac{\text{rate of effusion of A}}{\text{rate of effusion of B}} = \left(\frac{GMW_B}{GMW_A}\right)^{1/2}$$

Graham's Law has application in the determination of molecular weights as illustrated in Example 5.11 in the text, and for the calculation of relative molecular speeds of gases of known molecular weights.

Exercise 5.2

If the average speed of an oxygen molecule (O_2) is 4.45×10^4 cm per sec at a given temperature, what is the average speed of a nitrogen molecule (N_2) at the same temperature?

Exercise 5.2 Answer

Use Graham's law in the form $\dfrac{u_{N_2}}{u_{O_2}} = \left(\dfrac{GMW_{O_2}}{GMW_{N_2}}\right)^{1/2}$

$u_{N_2} = u_{O_2} \left(\dfrac{32.0}{28.0}\right)^{1/2} = u_{O_2} (1.14)^{1/2} = 4.45 \times 10^4$ cm/sec(1.07)

$= 4.76 \times 10^4$ cm/sec

3. Kinetic Theory of Gases

The postulates of the kinetic theory were developed to explain the behavior of gases. This model proposes that gases consist of independent molecules in continuous random motion undergoing elastic collisions with one another and with the walls of the container. The model also proposes that there is a distribution of molecular speeds. Associated with the distribution of molecular speeds is a distribution of energies related to the translational motion of the molecules. The average translational energy is directly proportional to the absolute temperature. Since the average translational energy is independent of the kind of molecule, at a given temperature the average energy is the same for any group of gas molecules. It is also proposed that for ideal gases, the volume of the molecules is negligible compared to the volume occupied by the gas, and that there are no attractive forces between molecules. Real gases do have attractive forces between molecules and the molecules have a finite volume. The van der Waals equation can be used in calculations involving real gases. The Ideal Gas Law, $PV = nRT = (1/3)(GMW)nu^2$, along with the appropriate values of R, can be used to calculate molecular speeds and molecular energies.

SUGGESTED ASSIGNMENT V

Text: Chapter 5

Problems: Numbers 5.1–5.3, 5.25, 5.27, 5.28, 5.30–5.33, 5.36, 5.38, 5.40

Solutions to Assigned Problems Set V

5.1

a. The number of moles of H_2 gas and the temperature do not change. So, $P_2V_2 = P_1V_1$ or $P_2 = P_1 \times \dfrac{V_1}{V_2} = 750 \text{ mmHg} \times \dfrac{600 \text{ cm}^3}{120 \text{ cm}^3}$
= 3750 mmHg

b. The pressure and the number of moles of CO_2 do not change. The volume is directly proportional to the absolute temperature (Charles' Law). Hence, $V_2 = V_1 \times \dfrac{T_2}{T_1}$. We need to change from the Celsius temperature scale to the absolute scale.

$10° + 273° = 283°K; 100° + 273° = 373°K$

Now, substituting numbers, we obtain:

$V_2 = 200 \text{ cm}^3 \times \dfrac{373°K}{283°K} = 264 \text{ cm}^3$

c. Here we use the Ideal Gas Law and solve for n. First, change to the absolute temperature scale.

$20° + 273° = 293°K$

$n = \dfrac{PV}{RT} = \dfrac{(15 \text{ atm})(24 \text{ liter})}{\left(0.0821 \dfrac{\text{liter atm}}{\text{mole}°K}\right)(293°K)} = 15 \text{ moles}$

d. We can solve this problem as in Example 5.5 in the text, or we can calculate the number of moles of CCl_4 in one liter at these conditions, then use the molecular weight to calculate the density.

$T = 100° + 273° = 373°K; P = 425 \text{ mmHg} \times \dfrac{1 \text{ atm}}{760 \text{ mmHg}} = 0.559$ atm

$$n = \frac{PV}{RT} = \frac{(0.559 \text{ atm})(1.00 \text{ liter})}{\left(0.0821 \frac{\text{liter atm}}{°K \text{ mole}}\right)(373°K)} = 1.83 \times 10^{-2} \text{ mole}$$

$GMW_{CCl_4} = 12.0 + 4(35.5) = 154$ g/mole

$(1.83 \times 10^{-2} \text{ mole})(154 \text{g/mole}) = 2.82$g

The density is 2.82g/liter.

5.2

The partial pressure of N_2 is equal to the mole fraction (X) of N_2 times the total pressure.

$$X_{N_2} = \frac{n_{N_2}}{n_{N_2} + n_{O_2}} = \frac{0.20}{0.20 + 0.10} = 0.67$$

$P_{N_2} = X_{N_2}(P_{total}) = (0.67)(1.00 \text{ atm}) = 0.67$ atm

5.3

a. Equation 5.13 is an expression for the average molecular speed, $u = \left(\frac{3RT}{GMW}\right)^{1/2}$

Since 3, R, and GMW are constant, $\frac{u}{2u} = \frac{(298)^{1/2}}{(T)^{1/2}}$ where T is the absolute temperature when the average speed is double the speed at 298°K.

$(T)^{1/2} = 2(298)^{1/2}$; $T = 298 \times 4 = 1192°K$

Increase the temperature to 1192°K or 919°C to double the average speed of the O_2 molecule. Since $E_{trans} = 3/2 RT$, the energy of motion will double when the temperture is increased to $2(298°K) = 596°K$.

b. Since the temperature is the same for all of the gases, the translational energy is the same for all. $E_{trans} = \frac{1}{2} GMW\, u^2$. The average speed will increase with decreasing molecular weight. The molecular weights are: N_2, 28; O_2, 32; Ar 39.9; H_2O, 18; CO_2, 44. The order of increasing average molecular speeds is $CO_2 < Ar < O_2 < N_2 < H_2O$.

5.25

Recognize that 48.0 cm = 480 mm

pressure unknown = pressure known + pressure due to Δh mmHg
= 746 mmHg + 480 mmHg = 1226 mmHg

5.28

This problem is similar to Example 5.3. Calculate the number of moles of CO_2, and use the Ideal Gas Law.

$$15.0g \times \frac{1 \text{ mole}}{44.0g} = 0.341 \text{ moles}; 0°C = 273°K$$

$$P = \frac{nRT}{V} = \frac{(0.341 \text{ moles})\left(0.0821 \frac{\text{liter atm}}{\text{mole °K}}\right)(273°K)}{20.0 \text{ liter}} = 0.382 \text{ atm}$$

5.30

This problem is similar to Example 5.5; we want to calculate the weight of one liter of H_2 at these conditions. Use Equation 5.8 in the form

$$d = \frac{(GMW)(P)}{RT} = \frac{(2.02 \text{g/mole})\left(\frac{740}{760} \text{ atm}\right)}{\left(0.0821 \frac{\text{liter atm}}{\text{mole °K}} (373°K)\right)} = 0.0642 \text{g/liter}$$

Since the two gases are at the same set of conditions, the number of moles of UF_6 will equal the number of moles of H_2. The densities are proportional to their molecular weights. The GMW of UF_6 = 238 + 6(19) = 352 g/mole.

$$\text{density } UF_6 = \text{density } H_2 \times \frac{\text{GMW of } UF_6}{\text{GMW of } H_2}$$

$$= 0.0642 \text{ g/liter} \times \frac{352}{2.02} = 11.2 \text{g/liter}$$

5.31

The determination of the simplest formula from percentage composition is similar to the calculations in Unit 3. Consider 100g of compound, then:

$$92.3\text{g C} \times \frac{1 \text{ mole C}}{12.0 \text{g C}} = 7.69 \text{ mole C}$$

$$7.7\text{g H} \times \frac{1 \text{ mole H}}{1.01 \text{g H}} = 7.62 \text{ mole H}$$

The simplest formula is CH. The GMW may be calculated directly as in Example 5.6. First express the variables in the proper units.

$$226 \text{ cm}^3 \times \frac{1 \text{ liter}}{1000 \text{ cm}^3} = 0.226 \text{ liter}; \quad 100° + 273° = 373°K;$$

$$755 \text{ mmHg} \times \frac{1 \text{ atm}}{760 \text{ mmHg}} = 0.993 \text{ atm}$$

$$\text{GMW} = \frac{\text{gRT}}{\text{PV}} = \frac{(0.573\text{g})\left(0.0821 \frac{\text{liter atm}}{\text{mole °K}}\right)(373°K)}{(0.993 \text{ atm})(0.226 \text{ liter})} = 78.2 \text{g/mole}$$

The formula weight of the simplest formula is 13.0. The molecular weight is about six times the formula weight. The molecular formula is C_6H_6.

5.32

The collected gas is a mixture of hydrogen and water vapor. The vapor pressure of water at 22°C is 20 mmHg.

Applying Dalton's Law: $P_{H_2} = P_{total} - P_{H_2O} = (740 - 20) \text{ mmHg} =$
720 mmHg = 720 mmHg × $\frac{1 \text{ atm}}{760 \text{ mmHg}}$ = 0.947 atm

Use the Ideal Gas Law to calculate the number of moles of H_2.

$$n = \frac{PV}{RT} = \frac{(0.947 \text{ atm})(2.50 \text{ liters})}{\left(0.0821 \frac{\text{liter atm}}{\text{mole °K}}\right)(295°K)} = 0.0978 \text{ mole}$$

Using conversion factors,

$$0.0978 \text{ mole } H_2 \times \frac{1 \text{ mole Zn}}{1 \text{ mole } H_2} \times \frac{65.4g}{\text{mole Zn}} = 6.40g$$

5.33

The equation for the reaction is

$$2H_2O(l) \rightarrow 2H_2(g) + O_2(g)$$

Since the gases are measured at the same temperature and pressure, the volume of H_2 is twice the volume of O_2.

5.36

a. The number of moles of each gas can be calculated using the Ideal Gas Law. For N_2: 200 ml = 0.200 liter

$$T = 25° + 273° = 298°K; \quad 750 \text{ mmHg} \times \frac{1 \text{ atm}}{760 \text{ mmHg}} = 0.987 \text{ atm}$$

$$n = \frac{PV}{RT} = \frac{(0.987 \text{ atm})(0.200 \text{ liter})}{\left(0.0821 \frac{\text{liter atm}}{\text{mole °K}}\right)(298°K)} = 8.07 \times 10^{-3} \text{ mole } N_2$$

For He; 300 ml = 0.300 liter; T = 125° + 273° = 398°K;
$$1200 \text{ mmHg} \times \frac{1 \text{ atm}}{760 \text{ mmHg}} = 1.58 \text{ atm}$$

$$n = \frac{PV}{RT} = \frac{(1.58 \text{ atm})(0.300 \text{ liter})}{\left(0.0821 \frac{\text{liter atm}}{\text{mole °K}}\right)(398°K)} = 1.45 \times 10^{-2} \text{ mole He}$$

b,c. The total pressure in the 1.00 liter container at 25°C will be the sum of the partial pressures exerted by each gas. We can calculate the pressure exerted by each gas independently when the volume and temperature are changed, and then add these partial pressures to obtain the total pressure. Another approach is to calculate the total pressure based on the total number of moles of gas present, and then calculate the partial pressures of each gas from the mole fractions.

Using the latter approach, total number of moles = 1.45×10^{-2} + 0.80×10^{-2} = 2.25×10^{-2} moles

T = 25° + 273° = 298°K; V = 1.00 liter

$$P = \frac{nRT}{V} = \frac{(2.25 \times 10^{-2} \text{ moles})\left(\frac{0.0821 \text{ liter atm}}{\text{mole °K}}\right)(298°K)}{1.00 \text{ liter}}$$

$$= 5.50 \times 10^{-1} \text{ atm}$$

$$X_{He} = \frac{1.45 \times 10^{-2}}{2.25 \times 10^{-2}} = 0.644; \quad X_{N_2} = 1.000 - 0.644 = 0.356$$

P_{He} = 0.550 atm × 0.644 = 0.354 atm

P_{N_2} = 0.550 atm × 0.356 = 0.196 atm

5.38

a. We can use Equation 5.13 to calculate the average molecular speed of each gas, or Equation 5.16 in the following form for a ratio of the average molecular speeds.

$$\frac{u_{N_2}}{u_{O_2}} = \left(\frac{GMW_{O_2}}{GMW_{N_2}}\right)^{1/2} = \left(\frac{32}{28}\right)^{1/2} = 1.07$$

The average speed of the N_2 molecules is 1.07 times that of the O_2 molecules.

b. Since the gases are at the same temperature, the energies of motion are the same.

c. The partial pressures are directly related to the mole fractions.

$$\frac{P_{N_2}}{P_{O_2}} = \frac{X_{N_2}}{X_{O_2}} = \frac{0.79}{0.21} = 3.76$$

The partial pressure of N_2 is 3.76 times that of O_2.

5.40

The times are inversely related to the rates of effusion. Substituting into Equation 5.18,

$$\frac{\text{time}_{O_2}}{\text{time}_{SO_2}} = \left(\frac{GMW_{O_2}}{GMW_{SO_2}}\right)^{1/2}; \quad \frac{100}{\text{time}_{SO_2}} = \left(\frac{32}{64}\right)^{1/2} = 0.71$$

$$\text{time}_{SO_2} = \frac{100 \text{ sec}}{.71} = 140 \text{ sec}$$

UNIT 6

The Electronic Structure of Atoms

The goal of this unit is to develop a model of the electronic structure of atoms that is consistent with experimental evidence.

INSTRUCTIONAL OBJECTIVES

6.1 Properties of Electrons in Atoms and Molecules
  1. Summarize the postulates of the quantum theory.

6.2 Experimental Basis of the Quantum Theory
  1. Account for the origin of atomic spectra.
  *2. Use Einstein's equation to calculate the energy change associated with the wavelength or frequency of a spectral line. (Probs. 6.1, 6.23)
  *3. Use Equation 6.4 to calculate the wavelengths of lines in the hydrogen atomic spectrum. (Prob. 6.24)

6.3 The Bohr Theory of the Hydrogen Atom
  1. Recognize that Bohr's theory introduced the idea of quantized energy levels as indicated by Equation 6.5.
  *2. Use Equation 6.5 and the Einstein equation to calculate the energy of transition or the wavelength associated with an electron transition in a hydrogen atom. (Prob. 6.2)
  *3. Calculate the energy in a ground state or the ionization energy when given the charge on the nucleus and appropriate constants and energy conversion factors. (Prob. 6.26)
  4. Summarize, qualitatively and in your own words, the Bohr model of the atom.

6.4 Waves and Particles
1. Recognize the dual particle-wave nature of light and of electrons.
2. Summarize the implications of the de Broglie relation?
*3. Use Equation 6.10 to calculate the allowable energies of a particle confined to a region of a given length. (Prob. 6.28)
4. Interpret the wave function as the probability of finding an electron at a point in space, and as being proportional to the electron charge density at that point.

6.5 Electron Arrangements in Atoms
*1. State the four quantum numbers that describe an electron, interpret the physical significance of these numbers, and apply the rules governing the assignment of them. (Probs. 6.5, 6.36)
*2. Relate quantum numbers to s, p, d, f notations.
3. State the Pauli exclusion principle.
4. State Hund's rule.
*5. Write the electron configuration when given the atomic number of an element. (Probs. 6.3, 6.31, 6.32, 6.34)
*6. Draw an orbital diagram when given the atomic number or the electron configuration of an element. (Probs. 6.4, 6.33)
*7. Sketch the spatial arrangement of s and p atomic orbitals.

6.6 Experimental Support for Electron Configurations
1. Define the term ionization energy.
2. Relate ionization energies to the electron configuration for a particular atom.

SUMMARY

1. Models of the Atom

Quantum theory proposes that energy is absorbed or released in certain sized units or packages called quanta. When a system changes from one allowed energy state to another, the package of energy that is absorbed or released must match the difference in energy between the two states.

Atomic spectra are explained by quantum theory as follows. Excited atoms, those which have absorbed energy and had their electrons move from the lowest energy level to a higher allowed energy level, are unstable and tend to return to a lower energy state. The energy released by the electron in the transition to a lower energy level may be in the form of light. The wavelength of the photon emitted is related to the energy change by the Einstein equation, $\Delta E = hc/\lambda$, where h is Planck's constant. (Recall that the wavelength, λ, of light is related to the frequency v, and speed, c, by the relationship $c = \lambda v$.)

Atomic spectra are experimental evidence that support the theory that electrons exist in different energy levels. Each kind of atom produces a spectrum containing a set of discrete lines, as compared to a continuous spectrum that contains essentially all wavelengths.

The model of the atom proposed by Niels Bohr was based on Rutherford's model of the nuclear atom and Planck's idea that the energy of radiation is proportional to the frequency, $E = h\upsilon$. Bohr imposed the restriction that an electron is only permitted to have certain energies. Electrons can make only certain allowed energy transitions by going from one energy level to another. The energy levels can be identified by a quantum number. The equation that Bohr derived for the hydrogen atom agreed with the observed atomic spectrum and Balmer's equation for the wavelengths in the visible hydrogen spectrum. However, his calculations did not agree with experimental evidence for systems with more than one electron. Bohr's model of electrons moving about the nucleus in paths or orbits of fixed radii was discarded.

The fact that electrons could be diffracted gave support to de Broglie's theory that particles have wave properties. The solution of equations in which the electron is treated as a wave gives a set of wave functions. Each wave function describes an atomic orbital with a characteristic energy that corresponds to a region in which there is a probability of finding an electron. The square of a wave function is interpreted as being proportional to the electron charge density at this point in space.

2. Arrangement of Electrons in Atoms

An electron can be identified by a set of four quantum numbers. Further, the Pauli exclusion principle requires that no two electrons in an atom can have the same set of four quantum numbers. Three of these numbers relate to the atomic orbital. The principal quantum number, n, indicates the principal energy level and can have positive integral values from one to infinity. The sublevels or subshells within a principal energy level are designated by the second quantum number ℓ. This number describes the shape of the orbital. There are n sublevels within a principal level of quantum number n, with integral values from 0 to $(n-1)$. The third quantum number m_ℓ is associated with the orientation of the electron cloud. Within a sublevel of quantum number ℓ, there are $2\ell + 1$ orbitals, with integral values from $-\ell$ to 0 to $+\ell$. The fourth quantum number m_s, the spin quantum number, is associated with the spin of the electron about its own axis. The maximum number of electrons in a given orbital is two, and the electrons must be of opposite spin. The values are $+\frac{1}{2}$ and $-\frac{1}{2}$. Refer to Table 6.2 for a summary of the allowed sets of quantum numbers.

The orbital occupancy by electrons is indicated by the electron configuration. Recall that the sublevels within a principal energy level are identified by the quantum number ℓ. The ℓ values of 0, 1, 2, 3 have the letter designations s, p, d, f, respectively. The number of electrons in a given subshell is indicated by writing the orbital designation and indicating the number of electrons with superscripts. The notation $2p^5$ indicates five electrons in the p orbitals of the second principal quantum level.

Orbital diagrams indicate the number and relative spins of electrons in each orbital. When using the Aufbau principle, the lowest energy orbitals fill first. For a set of orbitals of equal energy, the order of filling is such that as many electrons remain unpaired as possible, with their spins in the same direction. When electrons are paired in one orbital, the spins must be opposite.

The square of the wave function is proportional to the probability of finding an electron at points around a nucleus. The regions of high probability are dependent on the quantum numbers n, ℓ, and m_ℓ. The value of n is associated with distance from the nucleus. The shape of the s orbital is spherical. The p orbitals are sometimes described as being dumbbell shaped and are oriented along axes at 90° to each other.

Exercise 6.1
1. Write the electron configuration for the fluorine atom. The atomic number is 9.
2. Give the orbital diagram for the fluorine atom.
3. Write the electron configuration for the potassium atom. The atomic number is 19.
4. Give the set of four possible quantum numbers that describes the highest energy electron in the argon atom when the atom is in its first excited state.

Exercise 6.1 Answers
1. Always assume the atom is in the ground state unless a definite, excited state is indicated. The capacities of each level and sublevel are given in Table 6.3. Figure 6.10 indicates the relative energies of the sublevels. You must know the capacities and order of filling to write electron configurations. Nine electrons are required for fluorine. $1s^2, 2s^2, 2p^5$
2. Refer to the electron configuration for fluorine. Then draw, label, and fill the orbitals, using arrows to represent the electron spins.

$$\underset{1s}{\underline{\uparrow\downarrow}} \quad \underset{2s}{\underline{\uparrow\downarrow}} \quad \underset{}{\underline{\uparrow\downarrow}} \quad \underset{2p}{\underline{\uparrow\downarrow}} \quad \underline{\uparrow}$$

3. Nineteen electrons are required.

$$1s^2, 2s^2, 2p^6, 3s^2, 3p^6, 4s^1$$

4. The configuration in the ground state is

$$1s^2, 2s^2, 2p^6, 3s^2, 3p^6$$

A 3p electron, an electron with the greatest energy, would gain energy and move to the next highest energy level, 4s. The quantum numbers required to describe a 4s electron are $n = 4$, $\ell = 0$, $m_\ell = 0$, $m_s = +½$ or $-½$.

SUGGESTED ASSIGNMENT VI

Text: Chapter 6

Problems: Numbers 6.1–6.5, 6.23, 6.24, 6.26, 6.28, 6.31–6.34, 6.36

Solutions to Assigned Problems Set VI

6.1

a. The wavelength of the light that is emitted is equal to energy that is lost in going from higher to lower energy states, and only certain electronic energy states are allowed.

b. Use Equation 6.2 and proceed using the proper constants as in Example 6.1 in the text.

$$4014 \text{Å} \times \frac{1 \times 10^{-10} \text{m}}{1 \text{Å}} = 4.014 \times 10^{-7} \text{m}$$

$$E_{photon} = \frac{hc}{\lambda} = \frac{(6.626 \times 10^{-34} \text{joule sec})(2.998 \times 10^8 \text{m/sec})}{4.014 \times 10^{-7} \text{m}}$$

$$= 4.949 \times 10^{-19} \text{joules}$$

6.2

$$\Delta E = E_2 - E_1 = \frac{-B}{(2^2)} + \frac{B}{(1^2)} = \left(\frac{-2.179 \times 10^{-18}}{4}\right) + \left(\frac{2.179 \times 10^{-18}}{1}\right)$$

$$1.634 \times 10^{-18} \text{joules}$$

The conversion factor is obtained from Equation 6.3.

$$1.634 \times 10^{-18} \text{joules/atom} \times \frac{1 \text{ kcal/mole}}{6.95 \times 10^{-21} \text{joules/atom}} = 235 \text{ kcal/mole}$$

6.3

The atomic number is 7; there are 7 electrons. Using the Aufbau principle as illustrated in Example 6.5 in the text, $1s^2, 2s^2, 2p^3$

6.4

Use the electron configuration for the N atom and proceed as in

Example 6.6 in the text. Remember to apply Hund's rule in filling the p orbitals.

$$\frac{\uparrow\downarrow}{1s} \quad \frac{\uparrow\downarrow}{2s} \quad \frac{\uparrow \quad \uparrow \quad \uparrow}{2p}$$

The N atom should have three unpaired electrons.

6.5

The rules for assigning quantum numbers are illustrated in Table 6.2 and summarized as follows:

(a) n = principal quantum number, indicates the principal energy level, and has values of 1, 2, 3, 4, . . .

(b) ℓ = second quantum number, denotes the sublevels, and has values of 0, 1, . . . (n − 1). For an s electron, ℓ = 0; for a p electron, ℓ = 1; for a d electron, ℓ = 2; for an f electron, ℓ = 3

(c) m_ℓ = third quantum number, is associated with the orientation of the electron cloud, and has integral values of $+\ell$ . . . 0 . . . $-\ell$.

(d) m_s = spin quantum number, is associated with the spin of the electron about its own axis, and has values of $+\frac{1}{2}$ or $-\frac{1}{2}$.

Applying these rules to the electron configuration we wrote in Problem 6.3, we can assign quantum numbers as follows:

$1s^2$: 1, 0, 0, $+\frac{1}{2}$; 1, 0, 0, $-\frac{1}{2}$

$2s^2$: 2, 0, 0, $+\frac{1}{2}$; 2, 0, 0, $-\frac{1}{2}$

$2p^3$: 2, 1, 1, $+\frac{1}{2}$; 2, 1, 0, $+\frac{1}{2}$; 2, 1, -1, $+\frac{1}{2}$ (other m_ℓ and m_s values are possible)

6.23

This involves the Einstein equation in the form $\lambda = \dfrac{hc}{\Delta E}$. We need the value of Planck's constant, h, and the speed of light. The energy must be expressed in the proper units. Using conversion factors:

$$65.5 \text{ kcal/mole} \times \frac{6.95 \times 10^{-21} \text{ joules/particle}}{1 \text{ kcal/mole}} = 4.55 \times 10^{-19}$$

joules/particle

$$\lambda = \frac{hc}{E} = \frac{(6.626 \times 10^{-34}\,\text{joule/sec})(2.998 \times 10^{8}\,\text{meters/sec})}{4.55 \times 10^{-19}\,\text{joules}}$$

$$= 4.37 \times 10^{-7}\,\text{m}$$

$$4.37 \times 10^{-7}\,\text{m} \times \frac{1\,\text{Å}}{1 \times 10^{-10}\,\text{m}} = 4.37 \times 10^{3}\,\text{Å}$$

6.24

Substitute into $\lambda = 3646.00\left(\dfrac{n^2}{n^2 - 4}\right)$

$$3797.91 = 3646.00\left(\frac{n^2}{n^2 - 4}\right)$$

$$\frac{n^2}{n^2 - 4} = 1.04166;\ n^2 = 1.04166\,(n^2 - 4) = 1.04166 n^2 - 4.16664$$

$$0.04166 n^2 = 4.16664$$

$$n^2 = 100.0$$

$$n = 10.0$$

This is line number 8 in the Balmer series.

6.26

$E = \dfrac{-Z^2 B}{n^2}$ For Li, Z = 3 and in the ground state n = 1

$$E = \frac{-(3)^2 (2.179 \times 10^{-18}\,\text{joules})}{1^2} = -1.961 \times 10^{-17}\,\text{joules/ion}$$

Ionization energy $= 1.961 \times 10^{-17}\,\text{joules/ion} \times \dfrac{1\,\text{kcal/mole}}{6.95 \times 10^{-21}\,\text{joules/ion}}$

$= 2.82 \times 10^{3}\,\text{kcal/mole}$

6.28

In this problem we use Equation 6.10. The value of n = 1 for the minimum energy; h = 6.626×10^{-34} joule sec;

$d = 1 \times 10^{-14}$ meter; $m = \dfrac{1.00\,\text{g}}{1\,\text{mole proton}} \times \dfrac{1\,\text{mole proton}}{6.02 \times 10^{23}\,\text{protons}}$

$= 1.61 \times 10^{-24} \text{g} = 1.61 \times 10^{-27} \text{kg}$

$$\epsilon \min = \frac{n^2 h^2}{8md^2} = \frac{(1)^2 (6.626 \times 10^{-34})^2}{8(1.61 \times 10^{-27})(1 \times 10^{-14})^2} = 3.4 \times 10^{-13} \text{joule}$$

ϵ proton < binding energy of 1.3×10^{-12} joules

The minimum energy for a neutron is about equal to that of a proton since their masses are about equal. The binding energy is therefore greater than ϵ of a neutron. Both the proton and neutron would be held in the nucleus. The ϵ alpha is about equal to $\frac{\epsilon \text{ proton}}{4}$ since the mass of an alpha particle is about 4 times the mass of a proton.

$$\epsilon \text{ alpha} \approx \frac{3.4 \times 10^{-13} \text{joule}}{4} \approx 8.5 \times 10^{-14} \text{joule}$$

This is less than binding energy for nuclear particles and the alpha would be held in the nucleus.

6.31

The same principles used in Problem 6.3 and Example 6.5 in the text apply here.

a. Note that K^+ has 18 electrons. The electron configuration is the same as that of Ar. $1s^2, 2s^2, 2p^6, 3s^2, 3p^6$.

b. The F^- species has 10 electrons. The electron configuration is $1s^2, 2s^2, 2p^6$.

c. Al has 13 electrons. $1s^2, 2s^2, 2p^6, 3s^2, 3p^1$.

d. Fe has 26 electrons. Let [Ar] represent the electron configuration for the first 18 electrons. [Ar] $4s^2, 3d^6$.

6.32

a. Incorrect, this has too many 2s electrons.

b. Excited state, $2s^2$ electrons are in the 2p orbitals.

c. This is in the ground state.

d. Excited state, a 3s electron is in a 3d orbital.

e. Excited state, a 2p electron is in a 4f orbital.

f. Excited state, a 2s electron is in a 2p orbital.

6.33

Use the same principles as in Problem 6.4 and Example 6.6 in the text.

a. The electron configuration of Ti (22 electrons) is: $1s^2$, $2s^2$, 2^6, $3s^2$, $3p^6$, $4s^2$, $3d^2$

$\underset{1s}{\uparrow\downarrow}\ \underset{2s}{\uparrow\downarrow}\ \underset{2p}{\uparrow\downarrow\ \uparrow\downarrow\ \uparrow\downarrow}\ \underset{3s}{\uparrow\downarrow}\ \underset{3p}{\uparrow\downarrow\ \uparrow\downarrow\ \uparrow\downarrow}\ \underset{4s}{\uparrow\downarrow}\ \underset{3d}{\uparrow\ \uparrow\ \_\ \_\ \_}$

This atom would have two unpaired electrons, and 20 electrons that are paired.

b. The Sc atom has 21 electrons. The electron configuration is [Ar] $4s^2$, $3d^1$.

The orbital diagram is the same through the 3p orbitals as the Ti atom in (a). The 4s and 3d are filled as follows.

$\underset{4s}{\uparrow\downarrow}\ \underset{3d}{\uparrow\ \_\ \_\ \_\ \_}$

There is one unpaired electron and 20 electrons that are paired.

c. The Mn atom has 25 electrons. The electron configuration is [Ar] $4s^2$, $3d^5$. The 4s and 3d orbitals are:

$\underset{4s}{\uparrow\downarrow}\ \underset{3d}{\uparrow\ \uparrow\ \uparrow\ \uparrow\ \uparrow}$

There are five unpaired electrons and 20 paired electrons.

6.34

Since this is an ion with a +2 charge, two additional electrons would be required for the neutral atom. The ion has 16 electrons; the two additional electrons would give an atom with 18 electrons. This is Ar. The ion is Ar^{2+}. The electron configuration of the ion is $1s^2$, $2s^2$, $2p^6$, $3s^2$, $3p^4$.

6.36

The following could not occur:

b. 3, 0, −1, +½: If $\ell = 0$, m_ℓ must equal zero.

c. 2, 2, 2, 2: If n = 2, ℓ can equal either zero or 1. m_ℓ can only equal integral values of $-\ell \ldots 0 \ldots +\ell$, m_s can only equal +½ and −½.

d. 1, 0, 0, 0: m_s can be either +½ or −½ but not zero.

e. 2, −1, 0, ½: If n = 2, ℓ can equal zero or 1 but not −1.

f. 2, 0, −2, ½: If $\ell = 0$, $m_\ell = 0$.

UNIT 7

Periodic Table

The goal of this unit is to associate periodic relationships and properties of the elements with electron configurations, and to classify and predict physical and chemical properties of pure substances based on these relationships.

INSTRUCTIONAL OBJECTIVES

7.1 Structure of the Periodic Table
1. Recognize that elements are arranged in order of increasing atomic numbers in horizontal periods and vertical columns called groups.
2. Compare the number of elements in the various periods.
3. List some physical properties of the alkali metals.
*4. Predict the formulas and write equations for reactions between alkali metals and water, Group VI A, and Group VII A elements. (Prob. 7.20)
5. Recognize that the halogens exist as stable, diatomic molecules in the elementary form.
6. List some physical properties of the halogens; recognize the trends in physical and chemical properties throughout the group.
*7. Predict the formulas and write equations for reactions between halogens and hydrogen, Group I A, and Group II A elements. (Prob. 7.20)
8. Recognize the unique properties of hydrogen.

7.2 Correlation with Electron Configuration
*1. Relate group numbers to the number of electrons in the outermost principal energy levels. (Probs. 7.1, 7.25)
2. Compare the physical and chemical properties of transition metals (B subgroups) in periods 4, 5, and 6 to those of the alkali metals.
3. Recognize that the similarity in properties of the lanthanides is

related to the filling of the 4f sublevels; in the actinides, to the filling of the 5f sublevels.

7.3 Trends in Atomic Properties
  1. Realize that, in general, atomic radii decrease from left to right across a period and increase from top to bottom within a group.
  2. Recognize that the size of an atom is the result of a balance between electron-electron repulsions and electron-nucleus attractions.
  3. Realize that, in general, ionization energies tend to increase from left to right across a period and decrease from top to bottom within a group, an inverse correlation with atomic radii.
  4. Relate values for electronegativities to relative tendencies to attract electrons.
  5. Recognize that, in general, electronegativities increase from left to right across a period and decrease from top to bottom within a group, an inverse correlation with atomic radii.
  6. List some physical properties of metals; associate metallic characteristics with low ionization energy.
  7. List some physical properties of nonmetals; associate nonmetallic characteristics with a tendency to gain electrons.
  8. List two unique properties of the noble gases; relate these properties to their electron configurations.
  9. List some general physical and chemical properties of the metalloids.

7.4 Predictions Based on the Periodic Table
  *1. Predict physical properties of elements when given corresponding properties of surrounding elements in the Periodic Table. (Prob. 7.2)
  *2. Predict relative values of properties of elements such as electronegativity, ionization energy, atomic radius, metallic character. (Probs. 7.3, 7.24)
  *3. Predict the formulas of binary and ternary compounds when given the formulas of analogous compounds formed by elements in the same groups of the Periodic Table. (Probs. 7.4, 7.29)

7.5 Sources of the Elements
  1. Correlate the position of an element in the Periodic Table with the source of the element.
  2. Describe the chemical and physical processes involved in freeing elements from ore deposits.
  3. Summarize the practices that may be used to conserve deposits of metallic ores.

SUMMARY

1. Periodicity of Properties Related to Electron Configurations of Atoms

The chemical properties of an atom are associated primarily with the electron configuration of the outermost shell. The Periodic Table lists the elements in order of increasing atomic number in horizontal rows called periods. The periods are arranged so that elements with the same outermost electron configuration fall in vertical columns called groups.

In general, ionization energy and electronegativity increase from left to right across a period and decrease from top to bottom in a group. The atomic radii and metallic character decrease across a period and increase from top to bottom in a group. These properties are associated with the net attractive forces between the outermost electrons and the positive nucleus.

The Periodic Table systematizes the chemical and physical properties of the elements and serves as a basis for predicting properties as illustrated in Exercise 7.1.

Exercise 7.1
1. For each of the following pairs, select the species that has the larger radius

    K, Ca; Ca, Sr; Ca, Ca^{2+}; Br, Br^-

2. For each of the following pairs, select the species with the higher first ionization energy.

    K, Ca; Ca, Sr

3. Estimate the melting point of strontium metal. The melting points in °C for Ca, Ba, Rb, and Y are 838, 714, 39, and 1509, respectively.

4. The formula of strontium chlorate is $Sr(ClO_3)_2$. Predict the formula of calcium bromate.

Exercise 7.1 Answers
1. K has a larger radius than Ca. The general trend is for size to decrease across a period. This trend is based on the fact that Ca has a greater nuclear charge than K and that each nucleus is attracting 4s electrons. Sr is larger than Ca. The general trend is for size to increase down a group. The increase in distance of the 5s over the 4s orbitals is greater than the effect of increase in nuclear charges. The Ca atom is larger than Ca^{2+}. The Ca^{2+} ion has two fewer electrons, giving the effect of the attractions of 20 protons

on 18 electrons. The Br⁻ ion is larger than Br. There are repulsions between 36 electrons with a nucleus containing 35 protons to attract the 36th electron.
2. Ca has a higher first ionization energy than K. Ca has a larger nuclear charge and a smaller diameter than K. Ca has a higher first ionization energy than Sr. The greater distance of a 5s electron in Sr predominates over the increased nuclear charge of Sr.
3. The average, based on horizontal neighbors, is:

$$\frac{Rb + Y}{2} = \frac{39 + 1509}{2} = 774$$

The average, based on groupd trends, is:

$$\frac{Ca + Ba}{2} = \frac{838 + 714}{2} = 776$$

Both averages are close to the accepted value of 768.
4. Both Sr and Ca are in the same group (2A), and Cl and Br are both in group 7A. Predict the formula to be $Ca(BrO_3)_2$.

2. Sources of the Elements and Metallurgy

Due to their chemical reactivities, most elements exist in combined form in nature. General correlations exist between the position of an element in the Periodic Table and the state in which it is found. In addition, general relationships are exhibited between the position of an element in the Periodic Table and the chemical composition of its principal ore.

The physical and chemical processes that are employed in extracting elements from natural deposits depend upon the purity of the deposit and the chemical composition of compounds in the ore. Metals of relatively low chemical activity can be freed from their compounds by thermal decomposition. Chemical reducing agents such as carbon and carbon monoxide can be used economically on metals of moderate activity. Metals of high reactivity require the use of very strong reducing agents, or electrolytic processes such as those described in Chapter 22.

SUGGESTED ASSIGNMENT VII

Text: Chapter 7

Problems: Numbers 7.1–7.4, 7.20, 7.24–7.26, 7.29, 7.32

Solutions to Assigned Problems Set VII

7.1

Polonium is in Group 6A in Period 6. It has six electrons in the sixth principal quantum level. The outer electron configuration is $6s^2, 6p^4$.

7.2

The average based on group trends is $\dfrac{Cl + I}{2} = \dfrac{-101°C + 113°C}{2} = 6°C$.

The average based on period trends is $\dfrac{Se + Kr}{2} = \dfrac{218°C + (-157°C)}{2} = 30°C$. The estimate based on group trends is close to the observed value.

7.3

Recall that, in general, metallic character and atomic radii decrease as we move to the right and increase as we move down the table. Ionization energy and electronegativity increase from left to right and decrease top to bottom. Phosphorus is in Group 5A, Period 3. Germanium is in Group 4A in Period 4. Germanium should be more metallic than phosphorus and have a larger atomic radius. Phosphorus should have a higher ionization energy and be more electronegative than germanium.

7.4

Indium is in the same group as aluminum; selenium is in the same group as sulfur. We can predict $In_2(SeO_4)_3$.

7.20

a. Refer to Equations 7.1 and 7.3 in the text.

$$2Cs(s) + 2H_2O(l) \rightarrow 2CsOH + H_2(g)$$

$$2Cs(s) + Cl_2(g) \rightarrow 2CsCl(s)$$

b. $Ba(s) + 2H_2O(l) \rightarrow Ba(OH)_2 + H_2(g)$

$Ba(s) + Cl_2(g) \rightarrow BaCl_2(s)$

c. Refer to Equations 7.10 and 7.11 in the text.

$$La(s) + 3H_2O(l) \rightarrow La(OH)_3 + 1.5H_2(g)$$

$$2La(s) + 3Cl_2(g) \rightarrow 2LaCl_3(s)$$

7.24

Apply the general trends summarized in Problem 7.3 and predict:

a. metallic character As > Se > S
b. ionization energy S > Se > As
c. electronegativity S > Se > As
d. atomic radius As > Se > S

7.25

N: $1s^2, 2s^2, 2p^3$; O: $1s^2, 2s^2, 2p^4$

One of the p electrons is paired in oxygen, while in nitrogen the three p electrons are in three separate orbitals. The lower ionization energy of nitrogen may be associated with the special stability of the set of half-filled p orbitals.

7.26

a. The "effective nuclear charge" increases across a period. As a result, outer electrons are attracted more strongly to the nucleus, decreasing the atomic radius.

b. Hydrogen has one outer s electron, similar to Li. However, it is also lacking only one electron to complete its outer level. This is similar to F, which also lacks only one electron to complete its outer level.

c. There is a difference of only one proton between Mg and Al. There is a much greater difference in nuclear charge from Ca (At. No. 20) to Ga (At. No. 31), which in turn has a greater effect on the outer electrons.

7.29

a. Ca and Mg are in the same group; Bi is in the same group as P. Predict the formula to be $Mg_3(BiO_4)_2$

b. FrBr, analogous to CsCl

c. Au_2SO_4, analogous to Ag_2SO_4

d. K_2WS_4, analogous to Rb_2CrO_4

7.32

a. Convert the sulfide to an oxide; reduce the oxide with CO.

$$2NiS(s) + 3O_2(g) \rightarrow 2NiO(s) + 2SO_2(g)$$

$$NiO(s) + CO(g) \rightarrow Ni(s) + CO_2(g)$$

b. $BaCO_3(s) \xrightarrow{heat} BaO(s) + CO_2(g)$

$$3BaO(s) + 2Al(s) \rightarrow 3Ba(s) + Al_2O_3(s)$$

UNIT 8

Chemical Bonding

The goal of the unit is to consider the nature and properties of interatomic forces called chemical bonds.

INSTRUCTIONAL OBJECTIVES

8.1 Ionic Bonding
1. Recognize that the transfer of electrons from atoms of metals with low ionization energy to atoms of highly electronegative nonmetals produces ions.
2. Realize that the outer, incomplete energy levels are involved in electron transfers and that, in many instances, the octet of a noble gas structure is produced as a result of the transfer.
3. Account for the fact that formation of an ionic solid from the elements is always an exothermic process by relating the sign and magnitude of the lattice energy to a hypothetical series of steps for the process.
4. Relate the charge on an ion to the electron configuration of the A group elements in the Periodic Table. (Prob. 8.1)
5. Recognize that the formation of cations from transition metals always involves the loss of outer s electrons. (Prob. 8.30)
6. List the trends in sizes of monatomic ions and atoms in the Group A elements.
7. Learn the names and formulas of the polyatomic ions listed in Table 8.3.
*8. Apply the rules discussed in Section 8.1 to name ionic compounds. (Probs. 8.1, 8.28)
*9. Predict the formulas, name the compounds, and write balanced equations for the preparation of ionic compounds from the elements when a Periodic Table is provided. (Prob. 8.29)

8.2 Nature of the Covalent Bond
1. Discuss the factors responsible for the energy minimum in covalent bond formation.
2. Recognize that according to the valence bond theory, a covalent bond consists of a pair of electrons of opposite spins filling an atomic orbital on both bonded atoms.

8.3 Properties of the Covalent Bond
1. Define the following terms: bond polarity, bond energy, bond length.
2. Predict relative bond polarities and the extent of ionic character based on differences in electronegativities. (Prob. 8.2)
3. Relate covalent bond lengths to atomic radii and the influence of the partial ionic character of the bond.
4. Realize that the bond energy is affected by the ionic character of the covalent bond.
5. Recognize that multiple bonds between the same two atoms result in increased bond strength and decreased bond length.

8.4 Lewis Structures; The Octet Rule
*1. Apply the appropriate rules to write Lewis structures for molecules and polyatomic ions. (Probs. 8.3, 8.38)
*2. Explain the concept of resonance; write reasonable structures for resonance forms. (Prob. 8.35)
3. State the octet rule and describe several exceptions to this rule.

8.5 Molecular Geometry
*1. Predict bond angles and molecular geometry from Lewis structures based on electron pair repulsion principles. (Probs. 8.4, 8.38)
*2. Predict whether a molecule will be polar or nonpolar, knowing or having derived its geometry. (Probs. 8.5, 8.40)

8.6 Hybrid Atomic Orbitals
1. List the types of hybrid atomic orbitals discussed in Chapter 8 and characterize each as to the method of formation, number of orbitals, and their orientation.
*2. Distinguish between sigma and pi bonds; identify sigma and pi bonds when given the formula of a molecule or polyatomic ion. (Prob. 8.43e)
*3. Predict the types of hybrid bonds present in a molecule or polyatomic ion given the Lewis structure or the geometry of the species. (Probs. 8.6, 8.43)

8.7 Molecular Orbitals
1. Describe the three basic operations in the molecular orbital approach.
*2. Write a molecular orbital diagram for simple diatomic species; interpret the diagram in terms of the number and energy of bonds

formed and paramagnetic properties of the species. (Probs. 8.7, 8.45)

SUMMARY

A chemical bond is an attractive force that holds atoms together. These attractive forces are electrical in nature. The outer shell, or valence electrons, is involved in the formation of chemical bonds. Lewis structures are a way of representing the valence electrons. Whether the atoms transfer electrons to form ionic bonds, or share electrons to varying degrees to form covalent bonds, the bonds form because there is a set of attractive forces that produces a state of minimum energy.

1. Ionic Bonds

The attractive forces between positive and negative ions are called ionic bonds. The transfer of electrons from an element of low electronegativity (a metal) to an element of high electronegativity (a nonmetal) results in the formation of ions. The formation of an ionic solid from the elements may be considered as a three step process with the energies of each step affecting the net release of energy for the overall process. The ions that are formed from the representative elements (A groups) have a stable, noble gas electron configuration of eight electrons in the outer shell. Transition metal ions have stable electron configurations that usually differ from the noble gas structures. (Table 8.2).

Electron transfer always results in a conservation of charge, the number of electrons donated is equal to the number of electrons accepted. This can be represented by using Lewis structures.

$$2Na\cdot \longrightarrow 2Na^+ + 2e^-$$

$$:\!\ddot{S}\!\cdot + 2e^- \longrightarrow \left[:\!\ddot{\ddot{S}}\!:\right]^{2-}$$

The overall representation of the production of sodium sulfide is

$$2Na\cdot + \cdot\ddot{S}\!: \longrightarrow 2Na^+ + S^{2-}$$

This implies that no molecules of sodium sulfide are formed; discrete charged particles result from the transfer of electrons. This is consistent with such properties of ionic compounds as high melting points and good electrical conductivities in the molten state and in water solution.

Learn the names and formulas of the common polyatomic ions listed in Table 8.3. When writing formulas of compounds, the ratio of positive to negative ions must be selected to give electroneutrality.

Exercise 8.1
1. Give the electron configuration of Ca, Ca^{2+}, Fe, Fe^{2+}, S, S^{2-}.
2. Write formulas for the following ionic compounds: ammonium phosphate, calcium nitrate, sodium carbonate, aluminum sulfate.

Exercise 8.1   Answers
1. Ca: [Ar] $4s^2$. The notation [Ar] represents the electron configuration of the noble gas core for argon. Ca^{2+} has the same electron configuration as argon, $1s^2, 2s^2, 2p^6, 3s^2, 3p^6$.

    Fe: [Ar] $3d^6, 4s^2$

    Fe^{2+}: [Ar] $3d^6$

    S: [Ne] $3s^2, 3p^4$

    S^{2-}: [Ne] $3s^2, 3p^6$

2. Ammonium phosphate: The formula is $(NH_4)_3PO_4$. The three NH_4^+ ions are required to balance the one PO_4^{3-} ion.

    The formula of calcium nitrate is $Ca(NO_3)_2$.

    The formula of sodium carbonate is Na_2CO_3.

    The formula of aluminum sulfate is $Al_2(SO_4)_3$

2. Covalent Bonds

A single covalent bond consists of a pair of electrons shared between two atoms. A covalent bond results when there is a relatively small difference in electronegativity between two atoms. For atoms of the same electronegativity, as in the formation of H_2 or F_2 molecules, the electron pair is shared equally and a nonpolar bond results.

Unequal sharing of the electron pair occurs when there is a difference in the electronegativities. This results in the formation of polar covalent bonds, with some ionic character. The displacement of electrons toward the more electronegative element gives an increased electron density or partial negative charge around the atom. A partial positive charge results on the less electronegative atom. The greater the difference in electronegativity, the greater the polarity, and the shorter and stronger the bond. Multiple bonds are stronger than single bonds between the same two atoms.

The electron distribution in molecules and polyatomic ions can be represented by Lewis structures. By sharing electrons in the valence shell, many atoms achieve a noble gas configuration of eight electrons, or an octet. The rules for writing Lewis structures are listed in Section 8.4 of the text.

In some cases, it is impossible to write a Lewis structure that obeys the octet rule and at the same time is consistent with the properties of the substance. Some species are best described as having fewer than eight electrons, while others require an excess of eight electrons. The concept of resonance is introduced to account for the fact that, in some instances, a single Lewis structure does not correctly describe the properties of a substance.

The bond angles and the geometry of covalently bonded species can be predicted based on the principle that the electron pairs surrounding an atom are oriented to be as far apart as possible. The orientation of electron pairs is summarized in Table 8.6 of the text. The electron pair repulsion principle also applies to species containing multiple bonds. Consider that, as far as geometry is concerned, a multiple bond behaves as if it were a single electron pair. Remember that the description of the structure of a species reflects the position of the nuclei rather than the position of the valence electrons. Thus the geometry of NH_3 is said to be pyramidal even though the orientation of the four pairs of electrons is tetrahedral.

Although a molecule may contain polar bonds, the molecule itself may not be polar. When the geometry of a molecule is such that polar bonds cancel one another because of a symmetrical arrangement, the result is a nonpolar molecule. Diatomic molecules with atoms of different electronegativities are always polar. When the geometry of a molecule does not produce a cancellation of polar bonds, the molecule is polar and exhibits a dipole moment.

Exercise 8.2
1. Draw Lewis structures for each of the following species: (The central atom is written first.) PCl_3, HF, ClO_2^-, NH_4^+, $SiCl_4$, N_2
2. Based on electron pair repulsion principles, predict the geometry of each of the above species.
3. Predict which of the above molecules will be polar.
4. Draw reasonable resonance structures for HCO_2^- (C is the central atom).

Exercise 8.2   Answers
1. PCl_3: The chlorines are arranged around the central P atom. The total number of valence electrons in this neutral species is the number contributed by P and three Cls or $5 + 3(7) = 26$. Subtracting the six electrons for the three covalent bonds leaves 20 electrons to be distributed as unshared pairs forming an octet around each atom.

:C̈l :P̈: C̈l:
   :C̈l:

HF: This is straightforward—seven valence electrons from F and one from H give a total of eight. Two are used for the covalent bond, leaving six to be distributed as unshared pairs around F. The H achieves the noble gas structure of He with only one pair of electrons, the pair in the H—F bond.

H :F̈:

ClO_2^-: As indicated, the central atom is Cl. Adding the charge of the ion to the total contributed by the chlorine and oxygen gives $7 + 2(6) + 1 = 20$. Subtracting 4 for the two bonds leaves 16 to be distributed as unshared pairs.

$$\left[:\ddot{\underset{..}{O}} : \ddot{\underset{..}{Cl}} : \ddot{\underset{..}{O}}: \right]^-$$

NH_4^+: The N is the central atom and the total number of electrons is $5 + 4(1) - 1 = 8$. One electron is subtracted to account for the positive charge. Distribution of electrons gives:

$$\left[\begin{array}{c} H \\ H :\ddot{N}: H \\ H \end{array} \right]^+$$

$SiCl_4$: Si is the central atom. The total number of valence electrons is 32. Eight electrons for the four bonds are subtracted leaving 24 to be distributed as unshared pairs.

$$\begin{array}{c} :\ddot{Cl}: \\ :\ddot{Cl}:\underset{..}{Si}:\ddot{Cl}: \\ :\ddot{Cl}: \end{array}$$

N_2: The distribution of 10 electrons in a way that does not violate the octet rule gives :N⋮⋮N:

2. The PCl_3 has four pairs of electrons surrounding the central atom, with one pair unshared. The geometry is described as a pyramid. The diatomic HF and N_2 molecules are linear. The ClO_2^- ion has four pairs of electrons surrounding the central Cl atom, with two unshared pairs. This leads to a bent (angular) molecule. The NH_4^+ ion and the $SiCl_4$ molecule both have four shared pairs of electrons and a tetrahedral geometry.

3. The polar bonds in PCl_3 do not cancel. PCl_3 should be polar. The polar bonds cancel in the tetrehedrally shaped $SiCl_4$. The molecule should be nonpolar. The bond in N_2 is not polar and the molecule is nonpolar. There is a difference in electronegativity between the atoms in the linear HF molecule. A polar molecule results.

4. Use the rules for writing Lewis structures. The two resonance forms would be:

$$\left[\begin{array}{c} \ddot{\text{O}} \\ \parallel \\ \text{H} - \text{C} \\ \diagdown \\ \ddot{\ddot{\text{O}}} \end{array} \right]^{-} \longleftrightarrow \left[\begin{array}{c} \ddot{\ddot{\text{O}}} \\ / \\ \text{H} - \text{C} \\ \diagdown\!\diagdown \\ \ddot{\text{O}} \end{array} \right]^{-}$$

(A line is used to represent a shared pair of electrons.)

3. Valence Bond Theory

Valence bond theory is based on wave mechanics and is actually an extension of the Lewis model of an electron pair bond. According to valence bond theory, a bond results when an atomic orbital of one atom overlaps an atomic orbital of another atom. The bonding orbital can contain one pair of electrons. Atomic orbitals that overlap to produce a symmetrical electron density along the internuclear axis form sigma bonds. An overlap of atomic orbitals that results in an unsymmetrical concentration of electron densities above and below the internuclear axis produces a pi bond.

While electron repulsion principles can be used to predict molecular geometries, concepts of hybridization of atomic orbitals must be utilized to "explain" geometries when employing the valence-bond model. The types of hybrid orbitals and their orientations are listed in Table 8.7. Note that in the formation of multiple bonds, one or more pairs of orbitals are not hybridized.

4. Molecular Orbital Theory

The concept of molecular orbitals is analogous to the concept of atomic orbitals, in that each molecular orbital is designated by its energy and the spatial distribution of its electron density. Molecular orbitals result from the combination of atomic orbitals as illustrated in Figure 8.12 in the text. Bonding orbitals concentrate the electron density between nuclei. Antibonding orbitals place the electron density outside the region between the nuclei and result in decreased stability of the molecule. Relative energies of molecular orbitals are also indicated in Figure 8.12. Electrons are distributed in molecular orbitals in much the same way as electrons fill atomic orbitals. The following relationship gives the number of bonds:

$$\frac{\text{number of bonding electrons} - \text{number of nonbonding electrons}}{2} =$$

number of bonds

UNIT 8—CHEMICAL BONDING

The molecular orbital model is useful in predicting the number of bonds, the number of unpaired electrons, and the paramagnetic properties.

Exercise 8.3
1. Indicate the hybridization of the carbon atom in each of the following: $CHCl_3$, $C_2H_2Cl_2$, C_2H_2
2. Using molecular orbital theory, predict the electronic structure for O_2 and O_2^+.
3. Determine the number of bonds in O_2 and O_2^+.

Exercise 8.3   Answers
1. $CHCl_3$: sp^3 hybrid with a tetrahedral geometry. $C_2H_2Cl_2$: sp^2 hybrid. This contains three sigma bonds, and a pi bond that results from the unhybridized p orbitals. C_2H_2: sp hybrid and two pi bonds.
2. The diagram for O_2 is given in Table 8.8 in the text. The O_2^+ would have one less $2\pi^*p$ electron than O_2.
3. O_2 has two bonds $\left(\dfrac{8-4}{2} = 2\right)$. The O_2^+ ion has 2½ bonds $\left(\dfrac{8-3}{2} = 2½\right)$..

SUGGESTED ASSIGNMENT VIII

Text: Chapter 8

Problems: Numbers 8.1–8.7, 8.28, 8.29, 8.30, 8.35, 8.38, 8.40, 8.43, 8.45

Solutions to Assigned Problems Set VIII

8.1

The formula of each compound is derived from the charges of the cation and anion and by application of the principle that the compound is electrically neutral.

a. The charges of the monatomic ions are related to the positions of the two elements, aluminum and sulfur, in the Periodic Table. Aluminum is in Group 3A; sulfur in 6A. The ions are Al^{3+} and S^{2-}. Two Al^{3+} and three S^{2-} are required to balance the charges. Formula: Al_2S_3

b. Referring to Table 8.3, the ions are NH_4^+ and SO_4^{2-}. The formula is $(NH_4)_2SO_4$.

c. Table 8.2 indicates that the Group 2B element zinc forms Zn^{2+}. Table 8.3 indicates that nitrate is NO_3^-. The formula is $Zn(NO_3)_2$.

8.2

The greater the difference in electronegativity between two elements, the more polar the bond. In general, electronegativity increases across a period and decreases down a group. Arsenic and selenium are in the same period; Se should be more electronegative than As. Antimony is below and to the left of selenium. For these pairs of elements, Sb-Se should have the greatest difference in electronegativity. Tellurium lies below and to the right of arsenic; therefore, we should predict that these two elements have nearly the same electronegativities. The As-Te pair should be the least polar.

8.3

The procedures for writing Lewis structures are described in Section 8.4 and illustrated in Example 8.3. Applying these principles here, we have:

a. $\left(:\ddot{O} - \ddot{C}l - \ddot{O}:\right)^{-}$

b. $\left(:\ddot{O} - N = \ddot{O}\right)^{-}$

c. $\begin{bmatrix} :\ddot{O}: \\ | \\ :\ddot{O} - P - \ddot{O}: \\ | \\ :\ddot{O}: \end{bmatrix}^{3-}$

8.4

Refer to Table 8.6, which summarizes the relationship between geometries and the number of electron pairs.

a. The Lewis structure for ClO_2^- indicates two lone pairs and two oxygen atoms bonded to chlorine. The geometry is described as bent with a predicted bond angle of 109°.

b. Recall that the multiple bond behaves as if it were a single electron pair as far as molecular geometry is concerned. Thus for NO_2^-, there are two oxygen atoms bonded to nitrogen with one lone pair around the nitrogen. The bond angle should be about 120° for this bent structure.

c. If there are four oxygen atoms bonded to one phosphorus atom with no lone pairs, the geometry should be tetrahedral.

8.5

The charge centers do not coincide in the bent structures of ClO_2^- and NO_2^-; both should have a dipole. The polar bonds should cancel in the tetrahedral PO_4^{3-}; it is not a dipole.

8.6

Refer again to the Lewis structure for ClO_2^-. The bond angle was predicted to be 109°. Hybridization of the s and the three p orbitals of Cl to give an sp^3 hybrid is consistent with the predicted angle based on electron pair repulsion principles. In NO_2^-, the single bond between N and O, the unshared pair, and one of the double bonds are hybridized to give sp^2 hybridization. In PO_4^{3-}, the four coordinate covalent bonds result from sp^3 hybridization of P.

8.7

There are 10 valence electrons to consider in NO^+. Using the operations discussed in Section 8.7 gives:

	b	*	b	b	*	*
	σ2s	σ2s	π2p	σ2p	π2p	σ2p
NO⁺	2	2	4	2		

There are 8 electrons in bonding orbitals and two electrons in antibonding orbitals. The number of bonds $= \dfrac{8-2}{2} = 3$

8.28

Write the formulas for each of the compounds using the same principles applied in Problem 8.1.

a. $(NH_4)_2 SO_4$: Three moles of ions ($2\ NH_4^+$; $1\ SO_4^{2-}$).
b. $Al(NO_3)_3$: Four moles of ions ($1\ Al^{3+}$; $3NO_3^-$).
c. $Cr_2(CO_3)_3$: Five moles of ions ($2Cr^{3+}$; $3CO_3^{2-}$).
d. Rb_2Se: Three moles of ions ($2Rb^+$; $1Se^{2-}$).
e. $MgCl_2$: Three moles of ions ($1Mg^{2+}$; $2Cl^-$).

8.29

Decide upon the correct formulas for the reactants and products and then the proper coefficients to conserve mass and charge.

a. $3Ca(s) + N_2(g) \longrightarrow Ca_3N_2(s)$
b. $2CsH(s) \longrightarrow 2Cs(s) + H_2(g)$
c. $Ca^{2+}(aq) + CO_3^{2-}(aq) \longrightarrow CaCO_3(s)$
d. $Ba(s) + Cl_2(g) \longrightarrow BaCl_2(s)$

8.30

a. The Cr^{2+} is formed from the Cr atom when it loses the single 4s electron and one of the 3d electrons. The configuration is [Ar] $3d^4$.
b. The Co atom loses two 4s and one 3d in forming Co^{3+}. The configuration is [Ar] $3d^6$.
c. The two 3s electrons are lost giving the neon configuration of $1s^2$, $2s^2$, $2p^6$ for Mg^{2+}.
d. The nitrogen atom gains three electrons to give the neon configuration as in c.

88 · UNIT 8—CHEMICAL BONDING

8.35

    a. The octet rule is not obeyed in this odd electron species. These are resonance forms of NO.

    b. These are not resonance structures, but a rearrangement of atoms to give two different molecules.

    c. Again, the atoms are arranged in a different way. These are not resonance forms of the same species.

8.38

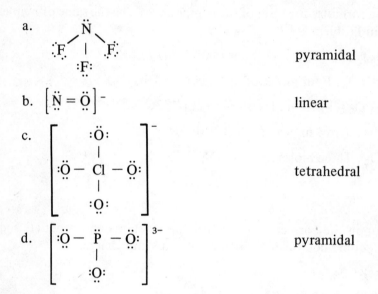

a. pyramidal

b. linear

c. tetrahedral

d. pyramidal

8.40

All except the ClO_4^- are dipoles.

8.43

    a. H_2O

    b. O_3

    c. BF_3

    d. BF_4^-

    e. N_2

8.45

The largest jump in ionization energy would be between the second and the third. The first two electrons are removed from the higher energy antibonding orbitals. The third electron is removed from a bonding orbital.

UNIT 9

Physical Properties as Related to Structure

The goal of this unit is to relate the physical properties of a pure substance to the nature of the structural units of the substance and the strength of the forces between these structural units.

INSTRUCTIONAL OBJECTIVES

9.1 Ionic Compounds
  1. Recognize that the relatively high melting points of ionic compounds are due to strong electrostatic forces between ions.
  2. Realize that ionic compounds in the molten state or in water solution are good electrical conductors.
  3. Recognize that ionic compounds tend to be soluble in polar solvents such as water but insoluble in nonpolar solvents.
  *4. Relate the melting points of ionic compounds to the charge densities of the ions (Probs. 9.1, 9.21)
  *5. Recognize that certain kinds of ionic compounds decompose upon heating. (Prob. 9.25)

9.2 Molecular Substances
  1. List the general physical properties of molecular substances.
  2. Distinguish between interatomic and intermolecular forces in molecular substances.
  *3. Relate trends in melting and boiling points of molecular substances to the molecular weights and polarity of these substances. (Probs. 9.2, 9.21)
  4. Describe the origin of intermolecular forces and compare the relative strengths of these forces.
  *5. Predict the types of intermolecular forces that will operate in a

variety of molecular substances. Correlate the physical properties of these substances with the types of forces. (Probs. 9.3, 9.23)

9.3 Macromolecular Substances
1. Recognize that a macromolecular substance is composed of a network of atoms held together by covalent bonds.
2. Relate the properties of macromolecular substances, such as allotropic forms of carbon and the silicates, to their structure and bonding.

9.4 Metals
1. List some properties that distinguish metals from other types of substances.
2. Interpret the properties of metals in terms of the electron-sea model.

SUMMARY

1. Ionic Compounds

The characteristic properties of ionic compounds, such as relatively high melting points, electrical conductivity in the molten state or in water solution, and solubility in polar solvents, can be related to the strong electrostatic forces between discrete, charged species. The magnitude of these attractive forces is dependent on the charge density of the ions as related by Coulomb's Law.

2. Intermolecular Forces

The attractive forces that hold particles of matter together are of several different kinds and of varying strengths. Since most of the different substances known to man exist in the form of molecules, the kind of attractive forces that are most frequently encountered are intermolecular forces. The magnitudes of these attractive forces determine such properties as melting point, boiling point, heat of fusion, and heat of vaporization. It is the attractive forces between the structural units (molecules), not the attractive forces within the structural unit (chemical bonds), that must be overcome to bring about a change in state of a molecular substance.

Intermolecular forces have relatively weak attractions, as indicated by the low melting and boiling points of molecular substances.

The types and origins of these forces, all electrical in nature, are described below:

A. Dipole-dipole. Due to unequal sharing of electrons and the molecular shape, the bonding electrons are not distributed symmetrically. This results in a charge separation that makes one end of the molecule slightly positive and the other end slightly negative, even though the whole molecule is electrically neutral. The intermolecular forces arise from the attraction between the positive end of one molecule and the negative end of another. These attractive forces account for the fact that, in general, polar molecules have higher melting and boiling points than nonpolar substances of similar molecular weights.

B. Hydrogen bond. This is an especially strong type of dipole attraction that arises between molecules in which hydrogen is bonded to small, highly electronegative atoms such as oxygen, fluorine, and nitrogen. The bond between hydrogen and the electronegative atom is highly polar and, in effect, makes the hydrogen behave almost as though it were a proton. This permits a strong electrical attraction between a hydrogen atom of one molecule with an oxygen atom in another molecule in a substance such as water. This type of attraction is exhibited between molecules of HF, HCN, and NH_3.

C. Dispersion forces. These forces are present in all matter and are the primary forces of attraction between nonpolar molecules. Dispersion forces result from the distortion or polarization of the electron cloud around a molecule. The instantaneous dipole that results from the polarization in one molecule induces a temporary dipole in a neighboring molecule. The larger the number of electrons in a molecule, the greater the distortion of the electron cloud and the stronger the dispersion forces. Since the number of electrons increases with increasing molecular weight, the strength of the dispersion forces usually increases with increasing molecular weights. This relationship is exhibited by the trend in melting points of the nonpolar molecules in Group 7. Molecular shape is also a factor in determining the strength of dispersion forces. Other factors being equal, the greater the surface available for contact between molecules, the larger the dispersion forces.

3. Macromolecular and Metallic Substances

Substances that contain a network of atoms joined to each other by covalent bonds are called macromolecular. Since no separate molecules exist, a crystal of such a substance may be viewed as one giant molecule. The allotropic forms of carbon and silicon dioxide are examples of macromolecular substances. Since these substances are held together by covalent bonds, crystals of this type have very high melting points.

The electron-sea model views a metal as positive ions in a sea of relatively few, loosely bound electrons that are free to move throughout the lattice structure. This model accounts for metallic properties such as high electrical and thermal conductivity and metallic luster. The variation in melting points among metals can be related to the magnitude of the charge on the positive ion.

Exercise 9.1
1. Based on the discussion of intermolecular forces, macromolecular substances, and the concepts of ionic, metallic, and covalent bonding, indicate the principal type of attractive force that must be overcome to:
   a. melt potassium
   b. boil NH_3
   c. melt KCl
   d. boil CCl_4
   e. melt diamond
2. Predict which member of each of the following pairs would have a higher melting point; in each case briefly explain your reasoning.
   a. Br_2 or Cl_2
   b. KCl or Br_2
   c. H_2O or H_2Te
   d. CCl_4 or CH_3Cl
   e. SiO_2 or CO_2

Exercise 9.1 Answers
1. a. Metallic bond, since K is a Group 1 metal.
   b. The principle attractive force is the hydrogen bond. Remember that hydrogen bonds are stronger than regular dipole-dipole attractions.
   c. Recognize that KCl is an ionic compound and that ionic bonds must be broken.
   d. CCl_4 is nonpolar; therefore, dispersion forces must be overcome to boil CCl_4.
   e. Diamond is an example of a macromolecular substance; therefore, covalent bonds must be broken to melt diamond.
2. a. Br_2. The dispersion forces in Br_2 are greater than in Cl_2 due to the greater number of electrons.
   b. KCl is ionic; therefore, electrostatic attractions must be overcome. In Br_2, only dispersion forces are involved.
   c. H_2O. Hydrogen bonding exists in H_2O. H_2Te exhibits weaker dipole-dipole attractions.
   d. CH_3Cl. Only dispersion forces are present in nonpolar CCl_4, while dipole attractions exist in the polar CH_3Cl molecule.
   e. SiO_2. Covalent bonds must be broken in a macromolecular substance such as SiO_2. The nonpolar CO_2 is held together by dispersion forces.

SUGGESTED ASSIGNMENT IX

Text: Chapter 9

Problems: Numbers 9.1–9.4, 9.18, 9.20, 9.21, 9.23, 9.25, 9.28

Solutions to Assigned Problems Set IX

9.1

The CaO should have a higher melting point than KCl. In CaO, the attraction is between +2 and −2 ions; in KCl, between +1 and −1 ions.

9.2

Cl_2 < Br_2 < ICl. ICl is polar while the nonpolar Br_2 has greater dispersion forces than Cl_2.

9.3

a. N_2, dispersion forces only.

b. H_2S, dipole forces as well as dispersion forces. (Remember, H_2S has an angular structure.)

c. H_2O, hydrogen bonding in addition to dispersion forces. (Oxygen is a relatively small, highly electronegative atom.)

9.4

The fact that the substance is a solid at room temperature, is soluble in water, and its water solution is a conductor, would indicate that it is an ionic compound. This eliminates substances (a) through (c). Since it decomposes to form another solid, we can conclude that the substance is $NiCO_3$.

9.18

a. Diamond has a three dimensional macromolecular structure. Graphite is a two dimensional structure with weak dispersion forces between the layers.

b. The dispersion forces are much greater in Cl_2 than in F_2 due to greater polarization of a larger electron cloud.

c. NaCl is composed of discrete ions that are mobile when fused.

d. Metals can absorb and re-emit light over a wide range of wavelengths. This is due to the fact that the outer electrons are not held tightly.

9.20

a. sugar–molecular

b. brass–metallic

c. chromium–metallic

d. propane–molecular

e. talc–macromolecular

9.21

a. MgO. The reasoning here is the same as that in Problem 9.1.

b. MgO. This is due to the smaller radius of Mg as compared to the radius of Ba.

c. NH_3. The NH_3 exhibits hydrogen bonding.

d. SbH_3. SbH_3 has a greater molecular weight than PH_3.

9.23

N_2. The HCl is polar, Cr metallic and $MgCO_3$ ionic.

9.25

b. $Al(OH)_3$ yields $Al_2O_3(s)$ and $H_2O(g)$

c. $CuCl_2 \cdot 5H_2O$ yields $H_2O(g)$ and $CuCl_2(s)$

d. $Al_2(CO_3)_3$ yields Al_2O_3 and $CO_2(g)$

9.28

If the substance is metallic, it will be an excellent electrical and thermal conductor and, when polished, will exhibit a metallic luster. An ionic substance may be soluble in water; no metal is water soluble.

UNIT 10

Introduction to Organic Chemistry

The goal of this unit is to apply principles of atomic structure, chemical bonding, and intermolecular forces to a study of the structure and properties of organic compounds.

INSTRUCTIONAL OBJECTIVES

10.1 Nature of Organic Substances
 1. Review the concepts of hybridization and the formation of single and multiple bonds for the carbon atom. (Chapter 8)
 2. Recognize that the unique ability of a carbon atom to form strong bonds with other carbon atoms results in a large number and variety of carbon containing compounds.
 *3. Relate the physical properties of organic substances to the types and magnitude of intermolecular attractions. (Prob. 10.42)

10.2 Hydrocarbons
 1. Recognize that alkanes (saturated hydrocarbons) contain only single carbon-carbon bonds formed from tetrahedrally oriented sp^3 hybrids.
 *2. Realize that some alkanes form ring structures (cycloalkanes). Draw structural formulas for these molecules (Prob. 10.31)
 *3. Recognize that the general formula of an alkane containing n carbon atoms is C_nH_{2n+2} and for cycloalkanes is C_nH_{2n}.
 *4. Write Lewis structues and structual formulas for the isomers of a hydrocarbon when given the molecular formula. (Prob. 10.1a)

*5. Relate physical properties of hydrocarbons to their structures and molecular weights.
*6. Write equations for the combustion of hydrocarbons.
*7. Describe the bonding and geometry in the unsaturated hydrocarbons known as olefins (alkenes).
*8. Draw structural formulas and label the *cis-trans* forms for the geometrical isomers when given the molecular formula of an olefin. (Probs. 10.1b, 10.25c)
*9. Describe the bonding and geometry in the series of unsaturated hydrocarbons known as acetylenes (alkynes). Draw structural formulas for these molecules (Probs. 10.1c, 10.27)
*10. Relate the chemical reactivity of alkenes and alkynes to the presence of double and triple carbon-carbon bonds.
*11. Describe the bonding and geometry of the benzene molecule.
*12. Draw structural formulas of the isomers of an aromatic hydrocarbon or its derivative when given the molecular formula.
*13. Classify a hydrocarbon as to type (alkane, alkene, alkyne, aromatic) when given or having derived the structural formula. (Prob. 10.29)

10.3 Sources of Hydrocarbons

1. List some of the fractions and their uses that are obtained from the distillation of petroleum.
*2. Summarize the processes used to increase the yield and quality of gasoline. (Prob. 10.2)
*3. Write equations for reactions involving synthesis gas and for the production of water gas. (Prob. 10.35)

10.4 Oxygen-Containing Compounds

*1. Classify an oxygen-containing organic compound as an alcohol, ether, aldehyde, ketone, acid, or ester when given the molecular formula. Draw structural formulas for these compounds. (Probs. 10.3, 10.25, 10.36, 10.37, 10.39)
2. Describe methods of preparation and the uses of methanol and ethanol.
3. Compare the chemical reactivities of ethers, aldehydes, and ketones.
*4. Write equations for the preparation of aldehydes and ketones from appropriate alcohols.
*5. Write equations for the preparation of organic acids and esters.
*6. Identify the following groups: alkyl, hydroxyl, carbonyl, carboxyl.
7. Distinguish between a soap and a detergent.

10.5 Some Halogen-Containing Organic Compounds

*1. Write equations for the preparation of alkyl halides from olefins and from alkanes. (Prob. 10.4)
2. List some properties and uses of halogenated hydrocarbons.

SUMMARY

1. Hydrocarbons

Carbon atoms have the unique property of being able to form very strong covalent bonds with other carbon atoms to produce long chain molecules or molecules in which one carbon atom is bonded to one, two, three or four other carbon atoms. Carbon atoms can also form double and triple bonds. For these reasons, plus the fact that carbon compounds can exhibit isomerism, the number of carbon compounds that can exist is extremely large.

The simplest organic compounds are the hydrocarbons. In the alkane (paraffin) series of hydrocarbons, each carbon atom forms a single bond with other carbon atoms or with hydrogen atoms. The carbon bond angles are the tetrahedral angles associated with sp^3 hybridization. Cycloalkanes are hydrocarbons that also contain single carbon-carbon bonds. These molecules commonly contain three to six carbon atoms arranged in a ring structure. As a group, the saturated hydrocarbons are relatively unreactive chemically. The physical properties of these nonpolar molecules are related to the magnitude of the dispersion forces that account for the intermolecular attractions.

The alkenes (olefins) have molecules that contain a carbon-carbon double bond. The carbon sp^2 hybridization results in a planar structure with bond angles of 120°. The double bond consists of one sigma and one pi bond. Due to the presence of the double bond, the alkenes are more reactive than the alkanes. The alkynes are even more reactive due to a carbon-carbon triple bond. The triple bond results from sp hybridization, and consists of one sigma and two pi bonds.

The hydrocarbons exhibit structural isomerism due to the branching of the carbon chains. With the unsaturated alkenes and alkynes, isomers also arise due to the different possible positions of the double or triple bond. Geometrical isomers exist among the alkenes owing to the lack of free rotation about the carbon-carbon double bond.

Benzene is an example of an aromatic hydrocarbon. This molecule has a six carbon atom planar ring structure with delocalized pi electrons above and below the plane of the molecule. In most reactions, the ring remains intact with groups being substituted for the hydrogen rather than being added as in the alkenes and alkynes. Many commercially important aromatic compounds are known. These all contain one or more benzene rings.

Petroleum, natural gas, and coal are our principal sources of hydrocarbons. In addition to being our main sources of energy, these substances are important raw materials for the production of a wide variety of products by the chemical industry. Coal gasification processes, such as the Lurgi process, appear to be short term solutions to the heating fuel shortage. Even though chemical research has resulted in an increase in the quality and quantity of gasoline from the available petroleum, the limited known reserves of petroleum present a problem in the near future.

Exercise 10.1
1. Classify each of the following as an alkane, alkene, alkyne, or an aromatic hydrocarbon.
   a. C_3H_8
   b. C_3H_6
   c. C_8H_{10}
   d. C_5H_8
2. Write structural formulas for the isomers of $C_3H_6Cl_2$.
3. Two structural isomers exist for the olefin with the molecular formula C_4H_8. The names, 1-butene and 2-butene, are based on the location of the double bond. Draw structural formulas for these two compounds.
4. Which isomer of butene can exhibit geometrical isomerism? Draw and label structural formulas for the *cis* and *trans* forms.

Exercise 10.1 Answers
1. For this exercise, it is probably useful to derive structural formulas. The general formulas, in which n = the number of carbon atoms, are:

   alkane, C_nH_{2n+2}; alkene, C_nH_{2n}; alkyne, C_nH_{2n-2}; aromatic, C_nH_{2n-6}.
   a. This fits the general formula for an alkane. The structural formula is
   $$H_3C-\underset{\underset{H}{|}}{\overset{\overset{H}{|}}{C}}-CH_3$$
   b. The structural formula is
   $$H-\underset{\underset{H}{|}}{\overset{\overset{H}{|}}{C}}-\overset{\overset{H}{|}}{C}=C\underset{H}{\overset{H}{{<}}}$$
   and fits the general formula of an alkene.
   c. A structural formula could be ⌬$-CH_2-CH_3$. Any others would also contain a benzene ring. This is an aromatic hydrocarbon.
   d. This fits the general formula for an alkyne. Structural formulas for several isomers can be drawn.
2. The three carbon atoms will form a chain structure since branching does not occur until there are at least four carbon atoms. We can start by placing a Cl on different carbon atoms. This gives two possible structures.

$$\begin{array}{c} H\;\;Cl\;\;H \\ |\;\;\;|\;\;\;| \\ H-C-C-C-Cl \\ |\;\;\;|\;\;\;| \\ H\;\;H\;\;H \end{array} \quad \text{and} \quad \begin{array}{c} H\;\;H\;\;H \\ |\;\;\;|\;\;\;| \\ Cl-C-C-C-Cl \\ |\;\;\;|\;\;\;| \\ H\;\;H\;\;H \end{array}$$

There are two possible structures with two Cls on one carbon atom.

$$\begin{array}{c} H\;\;H\;\;Cl \\ |\;\;\;|\;\;\;| \\ H-C-C-C-Cl \\ |\;\;\;|\;\;\;| \\ H\;\;H\;\;H \end{array} \quad \text{and} \quad \begin{array}{c} H\;\;Cl\;\;H \\ |\;\;\;|\;\;\;| \\ H-C-C-C-H \\ |\;\;\;|\;\;\;| \\ H\;\;Cl\;\;H \end{array}$$

Use ball and stick models to convince yourself that any additional structures are identical to one of the above.

3.

$$\begin{array}{c} H \quad\;\; H\;\;H\;\;H \\ \backslash\;\;\;\;\;|\;\;\;|\;\;\;| \\ C=C-C-C-H \\ /\;\;\;\;\;\;\;|\;\;\;| \\ H \quad\;\;\;\; H\;\;H \end{array} \qquad \begin{array}{c} H\;\;H\;\;H\;\;H \\ |\;\;\;|\;\;\;|\;\;\;| \\ H-C-C=C-C-H \\ |\;\;\;\;\;\;\;\;\;\;\;\;\;| \\ H\;\;\;\;\;\;\;\;\;\;\;\;\;H \end{array}$$

    1-butene   2-butene

4. The 2-butene, due to lack of rotation about the double bond, can have the H that is attached to the double bonded carbon in either of the following positions.

$$\begin{array}{c} H\;\;H\;\;H\;\;H \\ |\;\;\;|\;\;\;|\;\;\;| \\ H-C-C=C-C-H \\ |\;\;\;\;\;\;\;\;\;\;\;\;\;| \\ H\;\;\;\;\;\;\;\;\;\;\;\;\;H \end{array} \qquad \begin{array}{c} H\;\;H\;\;\;\;\;\;\;H \\ |\;\;\;|\;\;\;\;\;\;\;| \\ H-C-C=C-C-H \\ |\;\;\;\;\;\;\;\;\;\;|\;\;\;| \\ H\;\;\;\;\;\;\;H\;\;H \end{array}$$

    cis   *trans*

2. Derivatives of Hydrocarbons

Certain substituents or groups change the chemical characteristics of hydrocarbons. The substituting atoms or groups of atoms are known as functional groups. It is the functional group that primarily determines the chemical properties of a molecule. Physical properties of derivatives containing electronegative atoms such as a halogen or oxygen are related to the polarity of the molecule. Dipole attractions and hydrogen bonding in the derivatives account for the higher boiling points and greater solubility in water than that of hydrocarbons of corresponding chain length.

Various methods may be employed to prepare alkyl halides. For example, the direct replacement of H atoms in an alkane produces a mixture of polysubstituted compounds. The reaction of HX or X_2 (where X = F, Cl, Br, I) with an olefin forms a product by the addition of the halogen to a carbon-carbon double bond.

Alcohols have the general formula R—OH, where R is an alkyl group. Isomers of the alcohols exist depending on the location of the —OH functional group, Aldehydes (general formula $R-\overset{\underset{\displaystyle |}{H}}{C}=O$) result from the mild oxidation of an alcohol that has the hydroxyl group (—OH) on a terminal carbon atom. Ketones (general formula $R-\overset{\overset{\displaystyle O}{\|}}{C}-R'$) are produced on mild oxidation of alcohols that contain the hydroxyl group on a nonterminal carbon atom. Both ketones and aldehydes contain the carbonyl group, C=O.

Organic acids (general fromula $R-\overset{\overset{\displaystyle O}{\|}}{C}-OH$) can be prepared by the oxidation of aldehydes or the alcohols from which the aldehydes are derived. The carboxyl group (—COOH) is the functional group in organic acids. Esters (general formula $R-\overset{\overset{\displaystyle O}{\|}}{C}-O-R'$) can be produced by the reaction between an alcohol and an organic acid. Ethers (general formula R—O—R'), which are isomeric with alcohols, are less reactive chemically than the other oxygen derivatives described above. However, ethers are quite volatile, indicating very little hydrogen bonding as compared to alcohols, and the vapors are very flammable.

The structures and formulas of the oxygen-containing derivatives are summarized in Table 10.3 in the text. Note that alcohols and ethers are isomeric, as are aldehydes and ketones, and acids and esters.

Exercise 10.2
1. Classify each of the following oxygen-containing organic compounds.

   a. $H_3C-\overset{\underset{\displaystyle |}{H}}{\overset{\displaystyle |}{C}}-\overset{\underset{\displaystyle OH}{|}}{\overset{\displaystyle |}{C}}-CH_3$

   b. $H_3C-\overset{\overset{\displaystyle O}{\|}}{C}-H$

   c. $H_3C-CH_2-\overset{\overset{\displaystyle O}{\|}}{C}-CH_3$

d. $H_3C-CH_2-O-CH_3$

e. 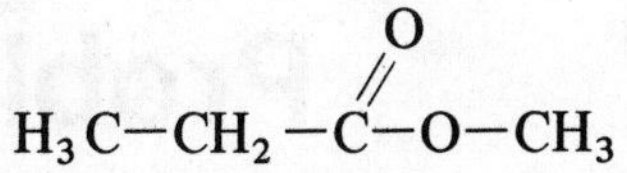 $H_3C-CH_2-\overset{\overset{O}{\|}}{C}-O-CH_3$

2. Write equations for the sequence of reactions beginning with methyl alcohol, CH_3OH, to produce (a) formaldehyde, H_2CO; (b) formic acid, $HCOOH$; (c) methyl formate, $HCOOCH_3$.

Exercise 10.2 Answers
1. If you know the general formulas and recognize the functional groups, this is easy. (a) alcohol, (b) aldehyde, (c) ketone, (d) ether, (e) ester.
2. The aldehyde can be formed by the controlled oxidation of methyl alcohol.

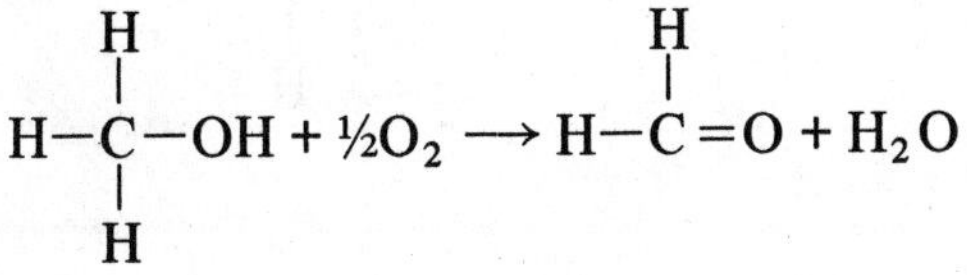

$$H-\underset{H}{\overset{H}{\underset{|}{C}}}-OH + \tfrac{1}{2}O_2 \rightarrow H-\overset{H}{\underset{|}{C}}=O + H_2O$$

Mild oxidation of formaldehyde yields formic acid.

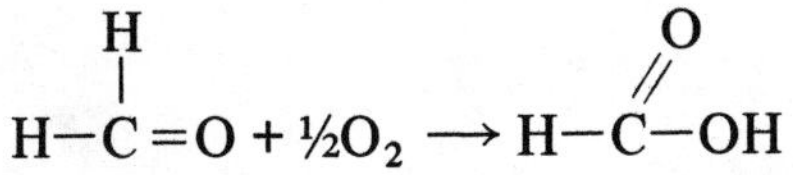

$$H-\overset{H}{\underset{|}{C}}=O + \tfrac{1}{2}O_2 \rightarrow H-\overset{\overset{O}{\|}}{C}-OH$$

The ester methyl formate can be prepared by reacting formic acid and methyl alcohol.

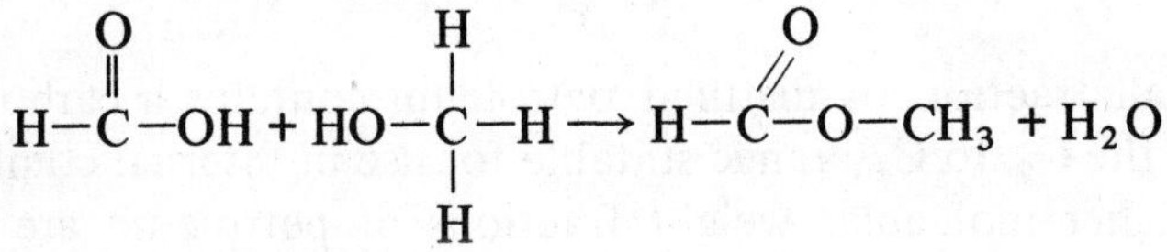

$$H-\overset{\overset{O}{\|}}{C}-OH + HO-\underset{H}{\overset{H}{\underset{|}{C}}}-H \rightarrow H-\overset{\overset{O}{\|}}{C}-O-CH_3 + H_2O$$

SUGGESTED ASSIGNMENT X

Text: Chapter 10

Problems: Numbers 10.1–10.4, 10.25, 10.27, 10.29, 10.31, 10.35–10.37, 10.39, 10.42

Solutions to Assigned Problems Set X

10.1

a. 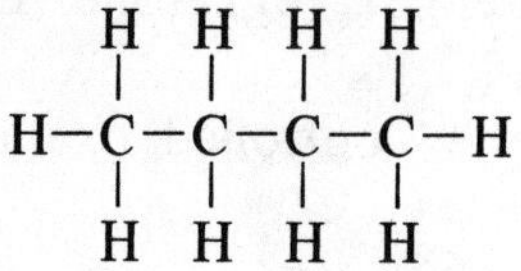 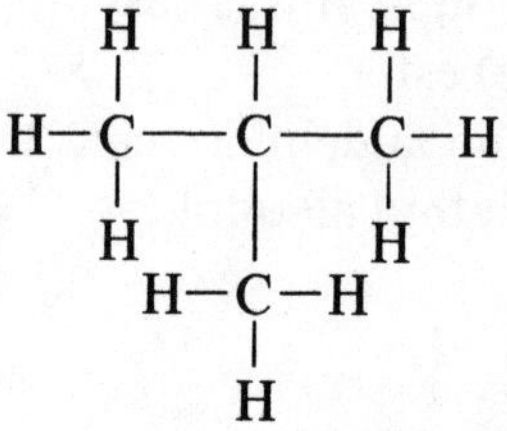
```
    H H H H              H   H   H
    | | | |              |   |   |
H―C―C―C―C―H          H―C――C――C―H
    | | | |              |   |   |
    H H H H              H   |   H
                           H―C―H
                             |
                             H
```

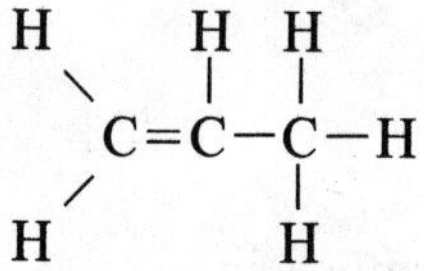

b.
```
 H      H H
  \     | |
   C=C―C―H
  /     |
 H      H
```

c.
```
          H
          |
H―C≡C―C―H
          |
          H
```

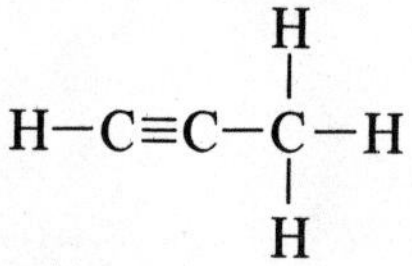

10.2

Only a small fraction of distilled petroleum contains a carbon atom content in the C_5 to C_{10} range suitable for use in internal combustion engines. Higher molecular weight fractions of petroleum are broken down to lighter fractions by a process known as catalytic cracking. Reforming or isomerization is used to produce branched chain molecules, which have better "antiknock" characteristics than straight chain molecules. Polymerization of light fraction molecules containing C_3 and C_4 olefins and paraffins to produce higher molecular weight paraffins also increases the yield of gasoline.

10.3

Refer to Table 10.3

a. 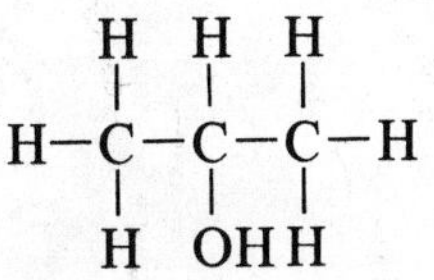
```
    H  H  H
    |  |  |
  H-C--C--C-H
    |  |  |
    H  OH H
```
This has the general formula $C_nH_{2n+2}O$ and the R—OH functional group. This is an alcohol.

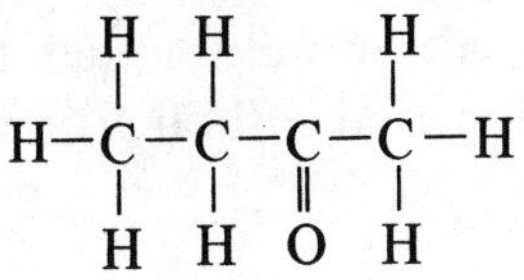

```
    H  H     H
    |  |     |
  H-C--C--C--C-H
    |  |  ‖  |
    H  H  O  H
```
The general formula here is $C_nH_{2n}O$ with the functional group R—C—R′.
                                    ‖
                                    O
This is a ketone.

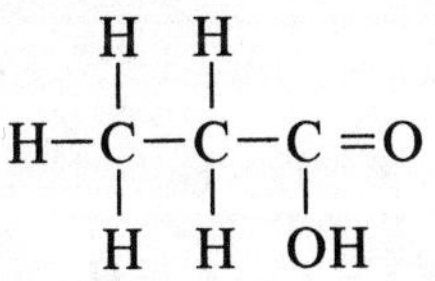

```
    H  H
    |  |
  H-C--C--C=O
    |  |  |
    H  H  OH
```
This is an acid. The general formula is $C_nH_{2n}O_2$ and an R—C—OH functional group.
                                                       ‖
                                                       O

b. The acid could have the molecular formula $C_2H_4O_2$.

10.4

$C_2H_4(g) + HBr(g) \rightarrow C_2H_5Br(l)$

$C_2H_6(g) + Br_2(g) \rightarrow C_2H_5Br + HBr(g)$

The latter reaction will produce a mixture of bromine substituted products that will require separation by distillation to obtain C_2H_5Br.

10.25

See Example 10.1 in the text. A line is used to indicate a bonding electron pair in the following structures.

a. 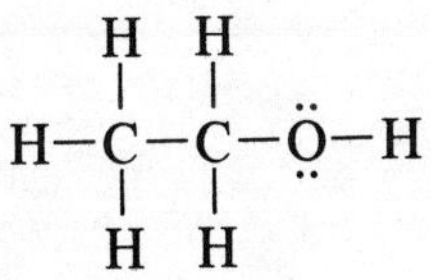
```
     H  H
     |  |
   H-C--C--Ö-H
     |  |
     H  H
```

b. 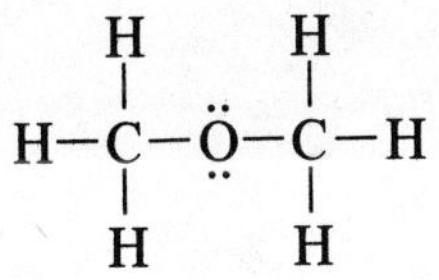
```
     H     H
     |     |
   H-C--Ö--C-H
     |     |
     H     H
```

c.
$$H-\underset{\underset{H}{|}}{\overset{\overset{H}{|}}{C}}-\underset{}{\overset{\overset{H}{|}}{C}}=\underset{}{\overset{\overset{H}{|}}{C}}-\underset{\underset{H}{|}}{\overset{\overset{H}{|}}{C}}-H$$

Note that *cis-trans* isomers exist as well as a structure that has the double bond located between the first and second carbon atoms.

10.27

This is a member of the alkyne series.

$$H-\underset{\underset{H}{|}}{\overset{\overset{H}{|}}{C}}-C\equiv C-\underset{\underset{H}{|}}{\overset{\overset{H}{|}}{C}}-H \qquad H-C\equiv C-\underset{\underset{H}{|}}{\overset{\overset{H}{|}}{C}}-\underset{\underset{H}{|}}{\overset{\overset{H}{|}}{C}}-H$$

10.29

		General formula	molecular formula
a.	paraffin	$C_n H_{2n+2}$	$C_6 H_{14}$
b.	olefin	$C_n H_{2n}$	$C_6 H_{12}$
c.	acetylene	$C_n H_{2n-2}$	$C_6 H_{10}$
d.	aromatic	$C_n H_{2n-6}$	$C_6 H_6$

10.31

The bond angle in an sp^2 hybrid is about 120°. The square structure with 90° angles for cyclobutadiene probably produces too great a strain on the bonds.

$$\begin{array}{c} H-C-C-H \\ \parallel \quad \parallel \\ H-C-C-H \end{array}$$

10.35

a. $4CO(g) + 2H_2(g) \rightarrow C_2H_4(g) + 2CO_2(g)$

b. $3CO(g) + 3H_2(g) \rightarrow C_2H_5OH + CO_2(g)$

10.36

a.
```
    H  H  H
    |  |  |
HO-C--C--C-OH
    |  |  |
    H OH  H
```

b.
```
  H      H  H  H
   \     |  |  |
    C=C--C--C-OH
   /     |  |
  H      H  H
```

c.
```
    H  O
    |  ||
 H-C--C-H
    |
    H
```

10.37

```
    H  H  H
    |  |  |
 H-C--C--C--C=O
    |  |  |  |
    H  H  H  H
      aldehyde
```

```
    H     H  H
    |     |  |
 H-C--C--C--C-H
    |  ||  |  |
    H  O  H  H
       ketone
```

10.39

a. This fits the general formula of C_nH_{2n+2}, an alcohol or an ether. The R—OH functional group is indicated in the formula. This is an alcohol.

b. The presence of the carboxyl group indicates that this is an acid.

c. This fits the general formula of C_nH_{2n} for an olefin (alkene).

d. The general formula is $C_nH_{2n}O_2$, that of an acid or an ester. The structural formula indicates the functional group of an ester.

10.42

The compound with the lowest boiling point is 2-methyl propane,
```
C-C-C
  |
  C
```
The electron cloud of a branched chain is not polarized as

greatly as that of a straight chain; therefore, it has smaller dispersion forces than butane, the compound with the next highest boiling point. Diethyl ether should be third. We predict that it has a boiling point close to that of a five carbon atom chain. Butyl alcohol has the highest boiling point due to hydrogen bonding.

UNIT 11

Liquids and Solids, Phase Changes

The goal of this unit is to consider the structure and properties of substances in the liquid and solid states and the transitions from one state to another.

INSTRUCTIONAL OBJECTIVES

11.1 Nature of the Liquid State
 1. Relate the properties of a liquid to the closeness of the molecules and the strength of attractive forces at short distances.
 2. Define molar heat of vaporization and relate this to the intermolecular forces in a liquid.

11.2 Liquid-Vapor Equilibrium
 1. Explain, on a molecular basis, liquid-vapor equilibrium.
 *2. Apply the gas laws and concepts of vapor pressure to calculate the volume or pressure of vapors in equilibrium and nonequilibrium conditions when appropriate data are given. (Probs. 11.1, 11.29)
 *3. Use Equation 11.3 to determine graphically the heat of vaporization of a liquid when given its vapor pressure at a series of temperatures. (Prob. 11.31)
 *4. Use the Clausius-Clapeyron equation (Equation 11.4) to calculate P_2, T_2, or ΔH_{vap} when given the values of two of these three quantities and the values of P_1 and T_1. (Prob. 11.2)
 5. Distinguish between the terms boiling point and normal boiling point.
 *6. Use Trouton's Rule to relate the molar-heat of vaporization of a liquid to its normal boiling point. (Prob. 11.33)

7. Explain what is meant by critical temperature and critical pressure and relate these to intermolecular attractions.

11.3 Nature of the Solid State
1. Relate the properties of a crystalline solid to the arrangement, type, and nature of the attractive forces between particles in the crystal lattice.
*2. Use the Bragg equation to calculate the interparticle distance when given the angle and order of diffraction for x-rays of known wavelength. (Prob. 11.35)
*3. Compare the number of atoms and their locations in simple, body-centered, and face-centered cubic cells.
*4. Relate the cell dimensions to such quantities as atomic radius and density when given the type of unit cell. (Probs 11.3)
*5. Distinguish between n-type and p-type semiconductors.

11.4 Phase Equilibria
*1. Construct and interpret a phase diagram for a pure substance when given the appropriate data; identify each area, line, and point on the diagram. (Probs. 11.4, 11.34)
2. Explain what is meant by melting point, triple point, sublimation, and heat of sublimation.
*3. Relate the heat flow for a given mass of a substance to the enthalpy changes associated with temperature and phase changes. (Prob. 11.5)

11.5 Nonequilibrium Phase Changes
1. Explain, on a molecular basis, superheating and supercooling.

SUMMARY

1. Liquids and Liquid-Vapor Equilibrium

The liquid state is characterized as consisting of molecules in random motion but close enough to one another so that intermolecular forces influence their physical behavior. For example, the closeness of molecules accounts for the fact that the volume of a liquid is much less sensitive to pressure and temperature changes than that of a gas, and that its density is much greater than that of a gas. The magnitude of the attractive forces operating over short distances accounts for the fact that each pure liquid has a characteristic surface tension, molar heat of vaporization, and equilibrium vapor pressure at a given temperature.

The pressure exerted by the molecules of a vapor in equilibrium with its liquid in a closed container varies with the temperature. As long as both liquid and vapor are present (i.e., if equilibrium can be established), the vapor pressure is independent of the volume of the container. However, if no liquid is present, the vapor behaves in accordance with the gas laws. Further, a vapor at a pressure greater than its equilibrium vapor pressure will condense to a liquid until the pressure decreases to the equilibrium vapor pressure. These principles are illustrated in Example 11.1 in the text.

The effect of temperature upon the vapor pressure is illustrated in Figure 11.5 in the text. The plot in Figure 11.5, along with Equation 11.3, illustrates a method of determining ΔH_{vap}. The slope can be determined by selecting any two points on the Y axis (Y_1 and Y_2) and reading the X axis values (X_1 and X_2) corresponding to these intersections.

$$\text{Slope} = \frac{\Delta Y}{\Delta X} = \frac{Y_2 - Y_1}{X_2 - X_1} = -2.4 \times 10^3$$

$$\text{Slope} = \frac{-\Delta H_{vap}}{(2.30)(1.99)} = -2.4 \times 10^3; \Delta H_{vap} = -(2.30)(1.99)(-2.4 \times 10^3) = 11,000 \text{ cal}$$

The Clausius-Clapeyron equation is useful in calculating the vapor pressure of a liquid at one temperature when the vapor pressure at another temperature and the heat of vaporization are known. This is illustrated in Example 11.2 in the text.

The boiling point is the temperature at which the vapor pressure of a liquid equals the external pressure. The normal boiling point is the temperature at which the vapor pressure equals 1 atm. The relationship between the heat of vaporization and the normal boiling point of a liquid is generalized by Trouton's rule. The critical temperature is the temperature above which the liquid phase of a pure substance cannot exist. At the critical temperature a gas can be liquified by application of the critical pressure.

Exercise 11.1
1. Predict which of the following liquids, hexane ($C_6 H_{14}$), heptane ($C_7 H_{16}$), or octane ($C_8 H_{18}$), will have the highest:
    a. ΔH_{vap}
    b. normal boiling point
    c. critical temperature
    d. equilibrium vapor pressure at 60°C
2. a. It requires 89.5 calories to vaporize 2.00g of liquid Br_2. Calculate the molar heat of vaporization of Br_2.
   b. Estimate the normal boiling point, in °C, of Br_2 using your value for ΔH_{vap} of Br_2.

Exercise 11.1 Answers

1. These four properties are related to the attractive forces between molecules. Dispersion forces are the only attractions in these essentially nonpolar molecules and are related to the molecular weight. Therefore, octane should exhibit the largest attractions and hexane the smallest. Predict that octane has the highest ΔH_{vap}, normal boiling point, and critical temperature. Since hexane has the least attractions, it should have the highest equilibrium vapor pressure at 60°C.

2. a. Using conversion units,

$$\frac{89.5 \text{ cal}}{2.00 \text{g}} \times \frac{160 \text{g}}{\text{mole}} = 7.16 \times 10^3 \frac{\text{cal}}{\text{mole}}$$

   b. Using Trouton's rule in the form

$$T_b \approx \frac{\Delta H_{vap}}{21} \approx \frac{7.16 \times 10^3}{21} \approx 341°K$$

Note that this gives the boiling point in °K. The boiling point in °C is 341° − 273° = 68°C. The measured value is 59°C.

2. The Solid State

Solids, in general, are characterized by highly ordered crystalline structures. As a result of this high degree of order, scientists have been able to investigate the structure of solids much more thoroughly than the structure of liquids. By measuring the angle at which x-rays of known wavelength are diffracted by a crystal, and using the Bragg equation, the distance between planes of atoms or ions can be calculated.

Characteristics of a crystal lattice can be described by the unit cell, the smallest unit which when moved a distance equal to the lengths of its edges generates the crystal lattice. If atoms occupy the crystal lattice points in a unit cell, only $\frac{1}{8}$ of a corner atom lies with a given cell, $\frac{1}{2}$ of a face-centered atom lies within a given cell, and an atom within the cell contributes the entire atom to the cell. The relations involved in the three types of unit cells are summarized below:

	simple	body-centered	face-centered
number of atoms/cell	1	2	4
atoms touch along	edge	body diagonal	face diagonal
relation between atomic radius (r) and cell length (ℓ)	$2r = \ell$	$4r = \ell\sqrt{3}$	$4r = \ell\sqrt{2}$

The above descriptions can be extended for crystals in which ions occupy the lattice points. These principles can be applied to calculate atomic and ionic radii. This type of calculation is illustrated in Example 11.3 in the text.

Most metals crystallize either in the face-centered cubic structure or in a body-centered cubic structure that contains only a slightly higher fraction of empty space. There are a variety of packing structures for ionic crystals due to different sizes of the two kinds of ions.

Imperfections that occur in crystals are called crystal defects. These defects affect the physical and chemical properties of a solid. The types of crystal defects, one of which has applications in the production of semiconductors, are discussed in the text.

Exercise 11.2

At very low temperatures, neon forms a crystal lattice in which the edge of a cubic unit cell is 4.52Å. The density of solid neon is $1.45 g/cm^3$.
1. Determine the number of neon atoms per unit cell.
2. What kind of unit cell is formed?

Exercise 11.2 Answers
1. This problem is similar to Example 11.3 in the text. Here you are given the density and asked to calculate the number of atoms per unit cell. Use the conversion factor method to convert Å to cm, calculate the volume of the cube, determine the mass of the cube, and solve for the number of Ne atoms that have this mass.

$$\left(4.52 \text{Å} \times \frac{1 \times 10^{-8} \text{cm}}{\text{Å}}\right)^3 \times \frac{1.45 g}{1.00 cm^3} \times \frac{6.02 \times 10^{23} \text{atoms Ne}}{20.2 g} =$$

4 Ne atoms

2. The unit cell that contains four atoms is the face-centered cubic cell.

3. Phase Equilibria

At the freezing point, a dynamic equilibrium exists between the solid and liquid phases. The energy that is required to change one mole of solid to a liquid is called the molar heat of fusion. The densities of most substances in the solid state are slightly greater than in the liquid state. The effect of pressure upon the melting point can be predicted by applying the principle that an increase in pressure favors the formation of the more dense phase. The process by which a solid changes to a vapor without passing through the liquid state is called sublimation. Solids exhibit a temperature dependent equilibrium vapor pressure.

The above characteristics of phase changes for solids and the previously discussed phase changes of liquids can be summarized by means of a phase diagram. The relationships between temperature, pressure, and phases for three different substances are illustrated in Figures 11.13 and 11.14 in the text. The essential features of these diagrams are summarized as follows. The three lines define the pressures and temperatures at which equilibria can exist between two phases. The liquid-vapor line, which terminates at the critical temperature and pressure, is the vapor pressure curve for the liquid. The solid-liquid line indicates the melting points at different pressures. The solid-vapor line is the vapor pressure curve for the solid. Only one phase can exist in a temperature-pressure region as delineated by these lines. Three equilibrium lines intersect at the triple point, the temperature and pressure at which all three states coexist in dynamic equilibrium.

SUGGESTED ASSIGNMENT XI

Text: Chapter 11

Problems: Numbers 11.1–11.5, 11.28, 11.29, 11.31, 11.33–11.35

Solutions to Assigned Problems Set XI

11.1

Convert the temperatures to °K and treat the vapor as an ideal gas.

$$300 \text{ mm Hg} \times \frac{323°K}{373°K} = 260 \text{ mm Hg}$$

A vapor at a pressure above its equilibrium vapor pressure is unstable. Condensation will occur and the vapor pressure will drop to the equilibrium vapor pressure of 92 mm Hg.

11.2

Convert the temperatures to °K. $50° + 273° = 323°K = T_1$; $P_1 = 272$ mm Hg; $T_2 = 80° + 273 = 353°K$; $P_2 = 760$ mm Hg; $R = 1.99$ cal/mole°K

Substituting into Equation 11.4:

$$\log_{10} \frac{760}{272} = \frac{\Delta H_{vap} (353 - 323)}{(2.30)(1.99)(353)(323)}$$

$$\log_{10} 2.79 = 0.446 = \frac{\Delta H_{vap} (30)}{5.22 \times 10^5}$$

$$\Delta H_{vap} = \frac{(0.446)(5.22 \times 10^5)}{30} = 7.76 \times 10^3 \text{ cal}$$

11.3

Refer to the sketch of the face of the unit cell in Example 11.3 in the text. For a face-centered cubic unit cell, atoms touch along a face diagonal, which has a length of $\ell\sqrt{2}$. Hence:

$$4r = \ell\sqrt{2}; \quad \ell = \frac{4r}{\sqrt{2}} = \frac{4(1.40 \text{Å})}{1.41} = 3.97 \text{Å}$$

11.4

a. The point at $-5°C$ and 10 mm Hg lies within the solid region: only ice is present.

b. At $-5°C$ and 1 mm Hg only vapor is present.

c. Point B corresponds to $25°C$ and 24 mm Hg and lies on the vapor pressure curve of the liquid. The two phases, liquid and vapor, are in equilibrium.

11.5

The value of ΔH for this process is related to the heat released when the gaseous water condenses at $100°C$, plus the heat released when liquid water cools from $100°C$ to $60°C$. Using the values of a ΔH_{vap} per gram and specific heat of liquid water,

$\Delta H = (2.00g)(-540 cal/g) + (2.00g)(1.00 cal/g°C)(60°C - 100°C)$

$= -11.60$ cal $= -1.16$ kcal

11.28

a. The weight of one mole of iron is 55.8g. Using the conversion factor approach: $55.8g \times \dfrac{1 cm^3}{7.87g} = 7.09$ cm^3

b. The formula for the volume of a sphere is $V = 4\pi r^3/3$ where r = the radius of the sphere. For one mole of iron atoms with an atomic radius of 1.26Å (1.26×10^{-8} cm): $V = (4/3)(3.14)(1.26 \times 10^{-8}$ cm$)^3 (6.02 \times 10^{23}) = 5.04$ cm^3

c. The amount of "empty space" is $7.09 cm^3 - 5.04 cm^3 = 2.05$ cm^3.

The fraction of "empty space" $= \dfrac{2.05 cm^3}{7.09 cm^3} = 0.289$

11.29

a. We can use the Ideal Gas Law in the form

$g = \dfrac{(GMW)PV}{RT}$ for this calculation. $GMW_{H_2O} = 18.0$ g/mole;

$V = 3.0 \times 10^4$ liters; $T = 298°K$. The pressure is the vapor pressure of water at saturation at 25°C. The pressure in atmospheres is 24/760 atm.

$$g = \frac{(18.0 \text{ g/mole})(24/760 \text{ atm})(3.0 \times 10^4 \text{ liters})}{\left(0.0821 \frac{\text{liter-atm}}{°K \text{ mole}}\right)(298°K)} = 7.0 \times 10^2 \text{ g}$$

b. If it requires 700g of water vapor to saturate the air and 800g are vaporized, the air is saturated and the pressure of the water vapor is the equilibrium vapor pressure of 24 mm Hg.

c. The air is not saturated, and equilibrium does not exist.

$$24 \text{ mm Hg} \times \frac{400 \text{g}}{700 \text{g}} = 14 \text{ mm Hg}$$

11.31

A plot of $\log_{10} P$ vs. $1/T$ gives a straight line with the slope = $\frac{-\Delta H_{vap}}{(2.30)(1.99)}$

For the temperature vapor pressure values:

T(°C)	20	30	40	50
P(mm Hg)	17.5	31.8	55.3	92.5
log P	1.243	1.502	1.743	1.966
1/T(°K)	0.003413	0.003300	0.003195	0.003096

$$\text{slope} \approx \frac{\Delta Y}{\Delta X} = \frac{Y_2 - Y_1}{X_2 - X_1} = \frac{1.966 - 1.243}{(3.096 - 3.413) \times 10^{-3}} = \frac{0.723}{-0.317 \times 10^{-3}} =$$

-2.28×10^3

$\Delta H = -(2.30)(1.99)(\text{slope}) = -(2.30)(1.99)(-2.28 \times 10^3) = 1.04 \times 10^4$ cal = 10.4 kcal

11.33

From Equation 11.5, ΔH_{vap} per mole = $21 \times T_b$. Note that T_b is in °K. Hence: ΔH_{vap} per mole = $21 \times 353 = 7400$ cal;

$$7400 \frac{\text{cal}}{\text{mole}} \times \frac{1 \text{ g}}{106 \text{cal}} = 70 \text{ g/mole}$$

11.34

a. True. The pressure is less than the vapor pressure of SO_2 at 25°C.

b. False. The pressure can not be greater than 3.8 atm at 25°C.

c. True. The vapor pressure at 150°C is less than 80 atm.

d. False. The vapor pressure of SO_2 reaches 1 atm at a temperature below 25°C.

11.35

Use Equation 11.6 in the form $\lambda = \dfrac{(2d)(\sin\theta)}{n}$. The sin 16°10' = 0.278; sin 33°50' = 0.556; sin 56°40' = 0.834

$$\lambda = \frac{2(2.76\text{Å})(0.278)}{1} = 1.53\text{Å}; \quad \lambda = \frac{2(2.76\text{Å})(0.556)}{2} = 1.53\text{Å};$$

$$\lambda = \frac{2(2.76\text{Å})(0.834)}{3} = 1.53\text{Å}$$

The wavelength was 1.53Å.

UNIT 12

Solutions

The goal of this unit is to develop an understanding of the factors that affect the solubility of a substance and to consider the physical properties of solutions.

INSTRUCTIONAL OBJECTIVES

12.1 Introduction
1. Distinguish between dilute and concentrated solutions.
2. Distinguish between solute and solvent.
3. Compare saturated, unsaturated, and supersaturated solutions.

12.2 Concentration Units
1. Learn the defining equations for the following concentration units: weight percentage, mole fraction, molarity, molality.
*2. Calculate the weight percentage of the components of a solution when given mass of the solute and mass of the solvent or total mass of the solution. (Probs. 12.1a, 12.29a)
*3. Calculate the mass or number of moles of solute and solvent when given the weight percentage of solute in a solution. (Prob. 12.28)
*4. Calculate the mole fraction of the components of a solution when given the number of moles of solute and solvent, or data from which the number of moles may be determined. (Probs. 12.1b, 12.29b)
*5. Calculate the molarity of a solution when given the number of moles of solute and the volume of the solution, or information from which the number of moles and the volume may be obtained. (Probs. 12.2a, 12.29d)
*6. Calculate the molality of a solution when given the number of moles of solute and the mass of the solvent, or information from which the number of moles and the mass may be obtained. (Probs. 12.1c, 12.29c)

*7. Given or having calculated any two of the following quantities, calculate the third: molarity, volume of solution, moles (or grams) of solute. (Probs. 12.28a, b)

*8. Relate the molarities and volumes of solutions when a second solution is prepared by dilution of the first solution. (Probs. 12.2b, 12.28c)

12.3 Principles of Solubility
1. Correlate the solubility of two liquids in each other with the similarity of their molecular structures and intermolecular forces.
2. Explain why the solubilities of different solids in a given liquid solvent are inversely related to their melting points.
3. Explain why the solubilities of different gases in a given liquid solvent are directly related to their normal boiling points.
*4. Predict the relative solubilities of two different solutes in a given liquid solvent, or the relative solubilities of a given solute in two different liquid solvents. (Probs. 12.34, 12.35)

12.4 Effect of Temperature and Pressure on Solubility
1. Recognize that an increase in temperature favors an endothermic process; predict the effect of an increase in temperature on the solubility of a solid or gas in a liquid solvent.
*2. Use Henry's Law to relate the solubility of a gas in a liquid solvent to the partial pressure of the gas. (Prob. 12.3)

12.5 Colligative Properties of Nonelectrolyte Solutions
1. Recognize that colligative properties of solutions depend primarily upon the concentrations of solute particles.
2. Compare a plot of vapor pressure vs. temperature for pure water to that for a water solution.
*3. Relate the vapor pressure of a solvent in a solution to the mole fraction of the solvent. (Prob. 12.4)
*4. Relate the total vapor pressure of a solution to the mole fractions of volatile solute and solvent. (Prob. 12.4)
*5. Relate the vapor pressure lowering to the mole fraction of the solute. (Prob. 12.36)
*6. Calculate the boiling and freezing points of a solution of a nonelectrolyte when given the molality of a nonvolatile solute (or information from which this may be calculated) and the necessary constants. (Probs. 12.5a, b)
7. Explain how osmotic pressure can be measured.
*8. Calculate the osmotic pressure when given the molarity of the solute. (Prob. 12.5c)
*9. Determine the molecular weight of an undissociated solute from measurements of vapor pressure lowering, freezing point depression, boiling point elevation, or osmotic pressure, when given weight and volume data and the necessary constants. (Probs. 12.6, 12.44)

SUMMARY

1. Solubility and Factors Which Affect Solubility

A solution is a homogeneous mixture of two or more substances. Although solutions may exist in the gaseous and solid states, our attention is focused on solutions in the liquid state. Solubility is a measure of the maximum amount of solute that dissolves in a given amount of solvent. For example, the number of grams of sugar that dissolve in one hundred grams of water is a measure of the "solubility" of sugar in water.

The solubility of a molecular substance is dependent on a number of factors. In general, liquid substances tend to be appreciably soluble in each other when they have similar molecular structures and when their intermolecular forces are of about the same magnitude. Thus water, in which hydrogen bonds are present, dissolves other hydrogen bonded molecules (e.g., ethyl alcohol) and nonpolar solutions dissolve other nonpolar molecules (e.g., hexane, C_6H_{14}, dissolves in heptane, C_7H_{16}). These two liquid pairs are completely miscible (they dissolve in all proportions). Solids, however, always exhibit a limited solubility in liquids. The closer a solid is to its liquid state (the lower the solid's melting point), the greater the solubility in a liquid solvent. Stated another way, low melting solids are more soluble than high melting solids of similar structure. The solubility of gases in liquids is also always limited. The closer a gas is to the liquid state (the higher its boiling point), the more soluble it is in a given liquid solvent. The "like dissolves like" generalization also applies to the solubility of solids and gases in liquids. The best solvent for a given gas or solid is one in which the intermolecular forces are similar to those of the solute.

Temperature and pressure also affect the solubility of a substance. The principle to apply here is that an increase in temperature increases the solubility if the dissolving process is endothermic. The solubility of most solids and liquids in a liquid solvent increases with increasing temperature. Many gases decrease in solubility with increasing temperature. Pressure has essentially no effect on the solubility of solids or liquids in liquid solvents. The solubility of a gas in a liquid is directly proportional to the partial pressure of the gas over the solution as related by Henry's Law.

Exercise 12.1
1. Predict which member of each of the following pairs of substances will be more soluble in benzene at 20°C and 1 atm. Benzene is nonpolar.
    a. C_2H_6 (g, bp = -88°C) or C_3H_8 (g, bp = -42°C)
    b. C_6H_{14} (l) or C_2H_5OH (l)
    c. CCl_4 or KI
    d. solid X (mp = 150°C) or solid Y (mp = 225°C); assume that X and Y have the same type of intermolecular forces.

122 • UNIT 12-SOLUTIONS

2. The solubility of O_2 gas in water is 4.42×10^{-2} g/liter at 1.00 atm and 20°C. At what pressure would 2.00×10^{-2} grams of oxygen dissolve in a liter of water at 20°C?

Exercise 12.1 Answers
1. a. C_3H_8. This alkane is closer to the liquid state (higher boiling point) than C_2H_6.
    b. C_6H_{14} is nonpolar, hence, it should be more soluble in the nonpolar benzene than the C_2H_5OH, which contains hydrogen bonds.
    c. CCl_4. Recognize that KI is ionic; CCl_4 is molecular.
    d. Solid X. It is closer to the liquid state (lower melting point).
2. Use Henry's Law; the solubility is directly proportional to partial pressure. As long as the concentrations are expressed in the same units, the relation is:

$$1.00 \text{ atm} \times \frac{2.00 \times 10^{-2} \text{g/liter}}{4.42 \times 10^{-2} \text{g/liter}} = 0.452 \text{ atm}$$

2. Concentration Units

The relative amounts of components of a solution may be described by the terms dilute and concentrated. The terms saturated, unsaturated, and supersaturated are also used to express relative concentrations. Be sure that you know the definitions for M, m, X, and % by weight for quantitatively expressing the concentrations of solutions.

In working with concentration units, keep the following in mind. Use the mass of the solution, not the mass of the solvent in weight percentage calculations. The total number of moles of all components, not just the number of moles of one component, appears in the denominator of the defining expression for the mole fraction. Be sure to distinguish between the number of moles and the number of moles per liter of solution when working with molarity. Finally, be sure to distinguish between molarity (M) and molality (m).

Exercise 12.2
1. Glycerol ($C_3H_8O_3$, Mol. Wt. 92.1) is the major component in "permanent" type antifreezes. A water solution of glycerol is prepared by adding 6.00 liters of glycerol (density 1.26g/ml) to 6.00 liters of water (density 1.00g/ml). Assume that the volumes of these two liquids are additive. Calculate for $C_3H_8O_3$ in this solution the:
    a. weight percentage
    b. mole fraction
    c. molarity
    d. molality
2. Water is added to 2.50 liters of the above glycerol solution to give

a second solution with a volume of 4.50 liters. Calculate the molarity of the second solution.

Exercise 12.2 Answers
1. a. Calculate the mass of each component, then apply the defining equation.

$$6.00 \times 10^3 \text{ ml} \times 1.26 \text{g/ml} = 7.56 \times 10^3 \text{ g } C_3H_8O_3$$

$$6.00 \times 10^3 \text{ ml} \times 1.00 \text{g/ml} = 6.00 \times 10^3 \text{ g } H_2O$$

$$\% \text{ glycerol} = \frac{7.56 \times 10^3 \text{ g}}{(7.56 \times 10^3 \text{ g}) + (6.00 \times 10^3 \text{ g})} \times 100 = 55.8\%$$

b. Calculate the number of moles of each component and then use the defining equation.

$$7.56 \times 10^3 \text{ g} \times \frac{1 \text{ mole}}{92.1 \text{ g}} = 82.1 \text{ moles } C_8H_8O_3$$

$$6.00 \times 10^3 \text{ g} \times \frac{1 \text{ mole}}{18.0 \text{ g}} = 333 \text{ moles } H_2O$$

$$X_{\text{glycerol}} = \frac{82.1}{82.2 + 33} = 0.198$$

c. Since the volumes of the components are additive, the volume of the solution is 12.0 liters.

$$M_{\text{glycerol}} = \frac{82.1 \text{ moles}}{12.0 \text{ liters}} = 6.84 \text{ moles/liter}$$

If the volumes were not additive, we would need to know the density of the solution so that the volume of the solution could be determined from the calculated mass of the solution in (a) above.

d. To calculate the molality, use the mass of water that was determined in part (a) of this exercise and the number of moles of $C_3H_8O_3$ from part (b).

$$\frac{82.1 \text{ moles}}{6.00 \text{ kg } H_2O} = 13.7 \frac{\text{moles}}{\text{kg } H_2O} = 13.7 \text{ m}$$

2. Remember that the number of moles of solute is not changed by dilution. Determine the number of moles of $C_3H_8O_3$ in 2.50 liters

of the original solution. This is the number of moles of glycerol in 4.50 liters of the second solution.

$$2.50 \text{ liter} \times 6.84 \frac{\text{moles}}{\text{liter}} = 17.1 \text{ moles}$$

$$\frac{17.1 \text{ moles}}{4.50 \text{ liters}} = 3.80 \text{ M}$$

3. Colligative Properties of Nonelectrolyte Solutions

When a solute is dissolved in a solvent, the vapor pressure of the solvent decreases. Raoult's Law relates the vapor pressure of the solvent in solution to the mole fraction of the *solvent* and the vapor pressure of the pure solvent. The freezing point of a solvent is also decreased by the addition of a solute. When the solute is nonvolatile, the boiling point of the solvent is elevated. Both the boiling point elevation and freezing point depression are due to the decrease in vapor pressure of the solvent. These effects are shown graphically in Figure 12.5 in the text.

Osmosis can also be explained in terms of vapor pressure lowering. The solvent moves through a membrane permeable only to the solvent from the region of high vapor pressure (pure solvent) to one of lower vapor pressure (solution). The movement of solvent molecules through the membrane into the solution may be prevented by applying pressure to the solution. The external pressure which is just sufficient to prevent osmosis is known as the osmotic pressure of the solution.

Each of the four effects described above is a colligative property. The magnitude of each effect is dependent primarily upon the concentrations of the solute particles rather than the kind of particles. Equations 12.7, 12.8, 12.9, and 12.10 in the text relate these effects to concentrations. These equations are also useful for the calculation of a molecualr weight of a solute as illustrated in Example 12.9 in the text.

Exercise 12.3
1. One of the reasons that glycerol is used as an antifreeze is due to that fact that it is a nonvolatile solute. Use the information in Exercise 12.2 to determine the vapor pressure of water in the glycerol-water solution when the temperature is 35°C (about 95°F). The vapor pressure of pure water at 35°C is 42.2 mm Hg.
2. Use the information in Exercise 12.2 to calculate the freezing point and boiling point, at 1 atm pressure, of the glycerol-water solution. For water, $k_f = 1.86$; $k_b = 0.52$
3. An aqueous solution containing 27.0g of glucose per 500 ml of solution may be used for intravenous feeding because it has the same osmotic pressure as blood. Calculate the osmotic pressure, in

atmospheres, of blood. Glucose, $C_6H_{12}O_6$, has a molecular weight of 180. Body temperature is 37°C.

4. When 1.000g of elemental sulfur is added to 38.50g of CCl_4, the freezing point of the CCl_4 is lowered by 3.25°C. The k_f for CCl_4 is 31.8. Determine the number of atoms of sulfur per molecule of sulfur.

Exercise 12.3 Answers

1. Calculate the mole fraction of water in the solution and employ Raoult's Law. Since the mole fraction of glycerol in Exercise 12.2 was calculated to be 0.198, the $X_{H_2O} = 1.000 - 0.198 = 0.802$

   $P = 0.802 (42.2 \text{ mmHg}) = 33.8 \text{ mmHg}$

2. The molality of the solution was calculated in Exercise 12.2 to be 13.7 m.

   $$\Delta T_f = k_f \times m = \frac{1.86°C}{m} \times 13.7m = 25.5°C$$

   The freezing point of the solution is $0.000°C - 25.5°C = -25.5°C$.

   $$\Delta T_b = k_b \times m = \frac{0.52°C}{m} \times 13.7m = 7.1°C$$

   The boiling point of the solution is $100.0°C + 7.1°C = 107.1°C$.

3. First calculate the molarity of the glucose solution.

   $$27.0g \times \frac{1 \text{ mole}}{180g} \times \frac{1}{0.500 \ell} = 0.300 \text{ M}$$

   Then use the relationship $\pi = MRT$. $R = 0.0821 \frac{\text{liter atm}}{°K \text{ mole}}$ and $T = 37° + 273 = 310°K$

   $$\pi = 0.300 \frac{\text{mole}}{\text{liter}} \times 310°K \times 0.0821 \frac{\text{liter atm}}{°K \text{ mole}} = 7.6 \text{ atm}$$

4. We need to determine the weight of a mole of sulfur molecules and compare this to the weight of a mole of sulfur atoms. Calculate the molality from the change in the freezing point.

   $$\Delta T_f = k_f \times m; \quad m = \frac{\Delta T_f}{k_f} = \frac{3.25}{31.8} = 0.102$$

   The number of grams of sulfur per kg of CCl_4 is

$$\frac{1.000 \text{g}}{0.03850 \text{ kg}} = 25.97 \text{g/kgCCl}_4$$

The 25.97g of sulfur is the weight of 0.102 moles of sulfur molecules. The weight of one mole of sulfur molecule is

$$\frac{25.97 \text{g}}{0.102} = 255 \text{g}$$

Since the atomic weight of S is 32.1:

$$\text{number of atoms per molecule} = \frac{255}{32.1} = 7.94$$

There are eight sulfur atoms per molecule of sulfur: the formula is S_8.

SUGGESTED ASSIGNMENT XII

Text: Chapter 12

Problems: Numbers 12.1–12.6, 12.28, 12.29, 12.34, 12.35, 12.36, 12.44

Solutions to Assigned Problems Set XII

12.1

a. $\% = \dfrac{gCH_3OH}{gCH_3OH + gH_2O} \times 100 = \dfrac{20.0g}{20.0g + 30.0g} \times 100 = 40.0\%$

b. Calculate the number of moles of CH_3OH and H_2O used in preparing the solution.

$20.0g \times \dfrac{1 \text{ mole } CH_3OH}{32.0g} = 0.625 \text{ mole } CH_3OH$

$30.0g \times \dfrac{1 \text{ mole } H_2O}{18.0g} = 1.67 \text{ mole } H_2O$

$X_{CH_3OH} = \dfrac{0.625}{0.625 + 1.67} = 0.272$

c. $m_{CH_3OH} = \dfrac{\text{number of moles } CH_3OH}{\text{kg } H_2O}$

The number of moles of CH_3OH was calculated in (b). Converting the weight of water to kg, $30.0g \times \dfrac{1 \text{ kg}}{1000g} = 0.0300$ kg, the molality of CH_3OH is calculated as follows:

$m = \dfrac{0.625}{0.0300} = 20.8$

12.2

a. Calculate the number of moles of $CaCl_2$:

$15.0g \times \dfrac{1 \text{ mole } CaCl_2}{111g} = 0.135 \text{ mole } CaCl_2$

The 0.135 mole is in 600 ml (0.600 ℓ) of solution.

The molarity is $\dfrac{0.135 \text{ mole}}{0.600 \text{ liter}} = 0.225 \dfrac{\text{mole}}{\text{liter solution}} = M$

b. In the final solution we want

$0.100 \dfrac{\text{moles}}{\text{liter}} \times 0.125 \text{ ℓ} = 0.0125$ mole of $CaCl_2$

We need to use enough 0.225 M $CaCl_2$ solution to give 0.0125 mole of $CaCl_2$. Using conversion factors:

$0.0125 \text{ mole} \times \dfrac{1 \text{ ℓ}}{0.225 \text{ mole}} = 0.0556$ ℓ of the 0.225 M $CaCl_2$ solution.

To prepare 125 ml of 0.100 M $CaCl_2$ from 0.225 M solution, add water to 55.6 ml of the 0.225 M solution until the total volume is 125 ml.

12.3

The solubility is directly proportional to the partial pressure of the gas phase over the solution.

6.90×10^{-4} mole/liter $\times \dfrac{0.79 \text{ atm}}{1.00 \text{ atm}} = 5.5 \times 10^{-4}$ mole/liter

12.4

The partial pressures are directly proportional to the mole fractions. The mole fraction of hexane is 0.40 and its vapor pressure is 122 mm Hg. $P_{hexane} = 0.40 \times 122$ mmHg = 49 mmHg

The mole fraction of heptane is 1.00 − 0.40 = 0.60.

$P_{heptane} = 0.60 \times 36$ mmHg = 22 mmHg

12.5

a. Before we can use Equation 12.9 to determine the freezing point lowering, we must calculate the molality of the glucose solution.

$$m = \frac{15.0}{180} \times \frac{1}{.200} = 0.417$$

$$\Delta T_f = k_f \times m = 1.86°C \times 0.417 = 0.776°C;$$

$$f_p = 0.00°C - 0.78°C = -0.78°C$$

b. Use $\Delta T_b = k_b \times m$ to calculate the boiling point elevation for the 0.417 m glucose solution.

$$\Delta T_b = 0.52°C \times 0.417 = 0.22°C$$

$$bp = 100.00°C + 0.22°C = 100.22°C$$

c. Assuming that $m \approx M$, use Equation 12.10.

$$\pi = MRT = 0.417 \frac{mole}{liter} \times 0.0821 \frac{liter\text{-}atm}{°K\ mole} \times 298°K = 10.2\ atm$$

12.6

The freezing point lowering is given as 4.10°C along with k_f for benzene. Using Equation 12.9 in the form

$$m = \frac{\Delta T_f}{k_f} = \frac{4.10°C}{5.10°C} = 0.804$$

$$\text{But, } m = \frac{g\ solute/GMW\ solute}{kg\ solvent} = \frac{10.0}{(GMW)(0.120)} = 0.804$$

$$\text{hence, GMW} = \frac{10.0}{0.120 \times 0.804} = 104 g/mole$$

12.28

a. From Table 12.1, we find that the weight percentage of HNO_3 is 72%. Therefore, $0.72 \times 100g = 72g$

b. The density is 1.42g/ml. Using the unit conversion approach:
$100\ ml \times 1.42g/ml \times 0.72 = 102g$

c. Dilute HCl is 6.0 M. The number of moles of HCl in one liter of 12M solution is:

$$1.00 \text{ liter} \times 12.0 \frac{\text{mole}}{\text{liter}} = 12.0 \text{ moles}$$

We want to determine the volume of 6.0 M HCl that will contain 12.0 moles of HCl.

$$12.0 \text{ moles} \times \frac{1.0 \text{ liter}}{6.0 \text{ moles}} = 2.0 \text{ liters of 6.0 M.}$$

We need to add 1.0 liter of H_2O to 1.0 liter of 12 M HCl solution to form a 6 M solution of HCl.

12.29

a. $\dfrac{30.0g}{30.0g + 50.0g} \times 100 = 37.5\%$

b. The number of moles of C_2H_5OH is

$$30.0g \times \frac{1 \text{ mole}}{46.0g} = 0.652 \text{ mole}$$

The number of moles CCl_4 is

$$50.0g \times \frac{1 \text{ mole}}{154g} = 0.325 \text{ mole}$$

$$X_{C_2H_5OH} = \frac{0.652}{0.652 + 0.325} = 0.667$$

c. $\dfrac{0.652 \text{ mole } C_2H_5OH}{0.0500 \text{kg}} = 13.0$

d. The mass of the solution is 80.0g. Using the density, we can calculate the volume of the solution.

$$80.0g \times \frac{1 \text{ ml}}{1.28g} = 62.5 \text{ ml} = 0.0625 \ell$$

$$\frac{0.652 \text{ mole}}{0.0625 \ell} = 10.4 \frac{\text{moles}}{\text{liter}}$$

12.34

We need to recognize that hexane is a nonpolar liquid.

a. The CH_3COOCH_3 is less polar than CH_3COOH. The magnitude of the intermolecular forces in CH_3COOCH_3 is closer to those of hexane; therefore, we predict CH_3COOCH_3 should be more soluble in hexane than CH_3COOH.

b. Ar is closer to the liquid state than He. The magnitude of the intermolecular forces in Ar is closer to those of hexane than He; predict Ar more soluble.

c. CCl_4 is nonpolar; NaCl is ionic. Predict that the nonpolar CCl_4 is more soluble in hexane.

d. The CH_3OCH_3 is less polar than CH_3OH. Predict CH_3OCH_3 for the same reasons as in (a).

12.35

The relative strengths of the dispersion forces are indicated by the fact that I_2 is a solid, Br_2 a liquid, and Cl_2 a gas at 20°C. The solubility is greatest for liquid bromine in liquid carbon tetrachloride, smaller for gaseous Cl_2 and solid I_2.

12.36

The vapor pressure lowering is related to the mole fraction of the solute by Equation 12.7, VPL = $X_2P_1°$. First calculate what the mole fraction, X_2, of the solute must be at 25°C.

$$X_2 = \frac{2.00 \text{ mm Hg}}{23.8 \text{ mm Hg}} = 0.0840$$

The mole fraction urea = $\frac{\text{moles urea}}{\text{moles urea} + \text{moles } H_2O}$

Let g = the number of grams of urea; the weight per mole is 60.0g. The weight of one mole of water is 18.0g. Since the mole fraction of urea is 0.0840, substitute into the above relationship:

$$0.0840 = \frac{g/60.0}{g/60.0 + 100/18.0} \quad ; g = 30.6g$$

At 50°C, $X_2 = \frac{2.00 \text{ mmHg}}{92.5 \text{ mmg}} = 0.0216$

As above, $0.0216 = \frac{g/60.0}{g/60.0 + 100/18.0} \quad ; g = 7.36g$

12.44

The molarity can be calculated from Equation 12.10 in the form $M = \frac{\pi}{RT}$. The osmotic pressure is 3.10/760 atm.

$$M = \frac{3.10/760 \text{ atm}}{\left(0.0821 \frac{\text{liter atm}}{°K \text{ mole}}\right)(293°K)} = 1.70 \times 10^{-4} \text{ mole/liter}$$

5.18g = 1.70 × 10⁻⁴ mole

1 mole = 30,500g

Assuming M ≈ m, substitute into the expression

ΔT_f = m × 1.86°C = 1.70 × 10⁻⁴ × 1.86°C = 3.16 × 10⁻⁴ °C;

T_f = −0.00032°C

UNIT 13

Water, Pure and Otherwise

The goal of this unit is to consider some special properties of water and electrolyte solutions. Problems associated with water pollution and treatment will also be discussed.

INSTRUCTIONAL OBJECTIVES

13.1 Water as a Solvent; Electrolyte Solutions
1. Relate the behavior of water as a solvent for ionic compounds to water's dielectric constant and hydration energy.
2. Distinguish between nonelectrolytes, weak electrolytes, and strong electrolytes.
*3. Relate the conductivities of solutions of strong electrolytes, at a given concentration, to the charge type. (Prob. 13.24)
4. Explain deliquescence.
*5. Determine the limiting value of the multiplier i in Equations 13.3 to 13.5 when given the formula of an electrolyte. (Prob. 13.1)
*6. Calculate ΔT_f, ΔT_b, or π for an electrolyte solution, given or having calculated the concentration and i. (Prob. 13.25)
*7. Calculate the nature or the extent of ionization of an electrolyte when given ΔT_f, ΔT_b, or π and the concentration, or data from which the concentration may be calculated. (Prob. 13.26)
8. Use the Debye-Huckel ionic atmosphere model to explain deviations from ideal behavior for electrolyte solutions.

13.2 Natural Sources of Water
1. List the principal ions found in seawater and compare with the principal ions found in river waters.

13.3 Water Pollution
1. Discuss the sources and types of water pollutants.
2. Explain how BOD is measured and what it means.
*3. Calculate the BOD of a water sample when given data for the amount of oxygen consumed by the sample and the amount of oxidizable substance in the sample. (Probs. 13.2, 13.31)
*4. Calculate the amount or concentration of an oxidizable substance in a water sample when given the BOD and the equation for the reaction with O_2. (Prob. 13.32)

13.4 Water Purification
1. Summarize the processes used in water treatment and purification.
2. Describe the undesirable properties of "hard" water.
3. Explain the chemical basis of the two common methods used to soften water.
*4. Determine the quantities of $Ca(OH)_2$ and Na_2CO_3 that should be added to soften water when given the concentrations of Ca^{2+} and HCO_3^- in the water sample. (Prob. 13.3)
5. Discuss various methods of desalination of water, and any problems associated with each method.

SUMMARY

1. Properties of Water and Electrolyte Solutions

Due to its unique properties, water has a great effect on the nature of our environment. Water's high specific heat and high heat of vaporization serve to control temperatures on the earth's surface. The fact that water attains its maximum density at 4°C is another special property that is responsible for mixing nutrients and dissolved oxygen in lakes. This also prevents bodies of water from freezing solidly from the bottom up. Water acts as a solvent for many ionic compounds owing to its high dielectric constant and the polar nature of its molecules, which leads to a high hydration energy.

The conductivities and colligative properties of electrolyte solutions are related to the concentration and charge type of the solute. Measurements of these properties can be used to determine the nature and extent of ionization of an electrolyte in solution. In solutions of electrolytes, ions are not totally independent of each other. As the concentration of an electrolyte increases, the ions influence each other to a greater extent, and as a result behave less independently. The Debye-Huckel ionic atmosphere theory, as well as several other proposed models, attempts to account for this nonideal behavior. The factor i is introduced into equations describing the colligative properties of electrolytes in order to take into account interactions between oppositely charged ions.

Exercise 13.1
1. Arrange the following 0.01 m water solutions in order of decreasing freezing points. Assume complete ionization. Li_2SO_4, $CsCl$, $Al_2(PO_4)_3$, $AlCl_3$.
2. The freezing point of a 3.0×10^{-3} molal water solution of the weak acid HA is $-0.0062°C$.

   The equation for the dissociation is

   $HA(aq) \rightleftharpoons H^+(aq) + A^-(aq)$

   a. Calculate the actual value of i at this concentration.
   b. What would the value of i be if HA were 100% dissociated? 0% dissociated?
   c. Calculate the % dissociation of HA in this solution.

Exercise 13.1 Answers
1. Since the value of m is constant in the expression $\Delta T_f = i(1.86°C)m$, all we need to do is evaluate i based on the formula of each ionic compound. Assuming complete ionization: i = 3 for Li_2SO_4; i = 2 for CsCl; i = 5 for $Al_2(PO_4)_3$; i = 4 for $AlCl_3$. The order of decreasing freezing points for these solutions is CsCl, Li_2SO_4, $AlCl_3$, $Al_2(PO_4)_3$.
2. a. Use the expression $\Delta T_f = i(1.86°C)(m)$ in the form

   $$i = \frac{\Delta T_f}{(1.86)(m)} = \frac{6.2 \times 10^{-3}}{(1.86)(3.0 \times 10^{-3})} = 1.1$$

   b. For 100% dissociation, i = 2; 0% dissociation, i = 1.
   c. Since the actual value of i is 1.1, the acid is about 10% dissociated.

2. Water Sources, Pollution, and Treatment

In general, the composition of dissolved substances in water is related to the natural source of the water. Ocean waters have a fairly constant composition of solute species, while lakes and rivers vary considerably in the types and concentrations of dissolved substances. Substances that directly or indirectly reduce the concentration of dissolved oxygen in water are among the most undesirable pollutants. The extent of such pollution can be measured by the BOD (biochemical oxygen demand) of a water sample. The types and sources of water pollutants are summarized in Table 13.5 in the text. Pollution due to detergents and insecticides is discussed in detail, along with effects of heavy metals and thermal pollution.

Treatment of municipal water supplies may involve sedimentation, coagulation, and filtration, depending upon the kinds and amounts of suspended materials in the source of the water. Disinfectants such as chlorine and ozone are used to kill disease-causing microorganisms in a water supply. Objectionable tastes and odors can be removed by filtration through beds of activated charcoal. When Ca^{2+} and Mg^{2+} ions are present in water they interfere with the cleansing action of soaps, and form deposits of carbonates and sulfates in water pipes and boilers. These ions can be removed by ion-exchange processes or by the lime-soda method.

Processes of desalination have been developed to provide fresh water in hot, arid regions. These processes include distillation, reverse osmosis, and ion exchange. Each of these methods has one or more economic or technologic problem associated with the process.

Exercise 13.2

1. An employee in a distillery accidentally released some ethyl alcohol into a stream that empties into a city reservoir. The BOD of the water in the reservoir increased by 48 mg/liter. Calculate the molarity of the alcohol in the reservoir. Assume the reaction is

$$C_2H_5OH(aq) + 3O_2(aq) \rightarrow 2CO_2(aq) + 3H_2O(l)$$

2. How many moles of $Ca(OH)_2$ and Na_2CO_3 should be added to soften ten liters of water in which the concentration of Ca^{2+} is 5.0×10^{-4} M and the concentration of HCO_3^- is 1.0×10^{-3} M.

Exercise 13.2 Answers

1. The equation indicates that three moles of O_2 are required to react with one mole of C_2H_5OH. Noting that 48 mg = 4.8×10^{-2} g, the molarity of alcohol is

$$\frac{4.8 \times 10^{-2} \text{g } O_2}{\ell} \times \frac{1 \text{ mole } O_2}{32 \text{g}} \times \frac{1 \text{ mole } C_2H_5OH}{3 \text{ moles } O_2} = 5.0 \times 10^{-4} \text{M}$$

(Note that this is only about 0.0023% alcohol by weight.)

2. The number of moles of Ca^{2+} and HCO_3^- in 10 liters of water is 5.0×10^{-3} and 1.0×10^{-2} moles, respectively. Adding 1 mole of $Ca(OH)_2$ for every 2 moles of HCO_3^- present,

$$1.0 \times 10^{-2} \text{ mole } HCO_3^- \times \frac{1 \text{ mole } Ca(OH)_2}{2 \text{ moles } HCO_3^-} = 5.0 \times 10^{-3} \text{ mole } Ca(OH)_2$$

The 5.0×10^{-3} mole $Ca(OH)_2$ also removes 5.0×10^{-3} mole of Ca^{2+} from the water; this is exactly the number of moles of Ca^{2+} originally present. No Na_2CO_3 needs to be added; only the 5.0×10^{-3} mole of $Ca(OH)_2$.

SUGGESTED ASSIGNMENT XIII

Text: Chapter 13

Problems: Numbers 13.1–13.3, 13.24–13.26, 13.31, 13.32

Solutions to Assigned Problems Set XIII

13.1

The complete dissociation of one mole of Na_3PO_4 would produce three moles of Na^+ and one mole of PO_4^{3-}. The limiting value of i would be four. To calculate i, use equation 13.3 in the form

$$i = \frac{\Delta T_f}{(1.86°C)m} = \frac{0.57°C}{(1.86°C)(0.10)} = 3.1$$

13.2

$$\text{The BOD} = \frac{\text{number mg } O_2 \text{ required}}{\text{number of liters sample}}$$

The problem actually asks how many milligrams of O_2 are required to react with 1.00g of benzene. Using conversion factors and the definition of BOD,

$$\frac{1.00\text{g benzene}}{1 \times 10^3 \text{ liters}} \times \frac{1 \text{ mole benzene}}{78.1 \text{ g benzene}} \times \frac{7.5 \text{ moles } O_2}{1 \text{ mole benzene}} \times \frac{32.0\text{g } O_2}{\text{mole } O_2} \times \frac{1 \times 10^3 \text{mg}}{\text{g}} = \frac{3.07\text{mg } O_2}{\text{liter}}$$

A BOD of 3.07 is almost within the acceptable range for "pure" water.

13.3

Add one mole of $Ca(OH)_2$ for every two moles of HCO_3^- present. The one mole of $Ca(OH)_2$ also removes one mole of Ca^{2+}. Since ten liters of water are to be treated, calculate the number of moles of Ca^{2+} and HCO_3^- present. Using conversion factors,

$$8.0 \times 10^{-4} \frac{\text{mole HCO}_3^-}{\text{liter}} \times 10 \text{ liters} \times \frac{1 \text{ mole Ca(OH)}_2}{2 \text{ moles HCO}_3^-} = 4.0 \times 10^{-3} \text{ mole Ca(OH)}_2$$

This removes 4.0×10^{-3} mole Ca^{2+} from the water. In ten liters of water there are:

$$6.0 \times 10^{-4} \frac{\text{mole Ca}^{2+}}{\text{liter}} \times 10 \text{ liters} = 6.0 \times 10^{-3} \text{ mole Ca}^{2+}$$

The number of moles of Ca^{2+} remaining is

$(0.0060 - 0.0040)$ mole $= 0.0020$ mole Ca^{2+}

It is necessary to add 2.0×10^{-3} mole of Na_2CO_3 to remove the remaining Ca^{2+}.

13.24

The principles here are to distinguish between nonelectrolytes, weak electrolytes, and strong electrolytes. Further, we need to recognize the number of moles of ions produced per mole of strong electrolyte.

The CH_3OH is a nonelectrolyte, HF a weak electrolyte, KNO_3 and $CaCl_2$ are both strong electrolytes. However, we predict that $CaCl_2$ will produce three moles of ions per mole of salt, while KNO_3 will produce only two. The order of increasing conductivities is $CH_3OH <$ HF $< KNO_3 < CaCl_2$.

13.25

The same principles that were applied in Problem 13.24 are used here to determine i in the equation $\Delta T_f = i \, (1.86°C)m$. Assuming ideal behavior, i = 1 for $CO(NH_2)_2$, 2 for KNO_3, 3 for $CaBr_2$, and 5 for $Al_2(SO_4)_3$.

For $CO(NH_2)_2$: $\Delta T_f = 1(1.86°C)(0.0010) = 0.00186°C$; $T_f = -0.0019°C$

For KNO_3: $\Delta T_f = 2(1.86°C)(0.0010) = 0.0037°C$; $T_f = -0.0037°C$

For $CaBr_2$: $\Delta T_f = 3(1.86°C)(0.0010) = 0.0056°C$; $T_f = -0.0056°C$

For $Al_2(SO_4)_3$: $\Delta T_f = 5(1.86°C)(0.0010) = 0.0093°C$; $T_f = -0.0093°C$.

13.26

We can calculate i using Equation 13.3 in the form

$$i = \frac{\Delta T_f}{(1.86°C)(m)} = \frac{0.024°C}{(1.86°C)(0.010)} = 1.3$$

The value of i would be one if HX did not ionize at all. If completely ionized, i would be two. The HX must be 30% ionized in a 0.010 molal solution.

13.31

The concentration of O_2 in pure water is 8.9 mg/liter. The amount of O_2 consumed is 6.2 mg/liter. The concentration of O_2 at equilibrium is (8.9 − 6.2) mg/liter = 2.7 mg/liter.

13.32

This is a problem in stoichiometry, in that we are asked to calculate the number of moles per liter of CN^- that reacts with 3.0 mg O_2 per liter. Using conversion factors,

$$\frac{3.0 \text{ mg } O_2}{\text{liter}} \times \frac{1 \text{ mole } O_2}{32 \times 10^3 \text{ mg}} \times \frac{1 \text{ mole } CN^-}{1.5 \text{ moles } O_2} = 6.3 \times 10^{-5} \frac{\text{mole } CN^-}{\text{liter}}$$

UNIT 14

Spontaneity of Reaction; ΔG and ΔS

The goal of this unit is to consider the factors that determine whether a reaction, at a certain temperature and pressure, will be spontaneous.

INSTRUCTIONAL OBJECTIVES

14.1 Criteria of Spontaneity; Useful Work
 1. Distinguish between the terms spontaneous and nonspontaneous.
 2. Realize that spontaneous processes are not necessarily rapid processes.
 3. Recognize that the criterion for the spontaneity of a process or reaction is the ability to produce useful work.

14.2 Free Energy Change, ΔG
 *1. Calculate $\Delta G^{1\ atm}$ for a reaction at 25°C, given a table of standard free energies of formation. (Probs. 14.1, 14.27, 14.33)
 2. Relate the sign of $\Delta G^{1\ atm}$ to the spontaneity of a reaction at 1 atm pressure.
 3. Recognize that $\Delta G = 0$ at equilibrium.

14.3 Entropy Change, ΔS
 1. Recognize that entropy is a measure of the disorder of a system.
 *2. Predict the sign of ΔS for physical and chemical processes. (Probs. 14.3, 14.30)
 *3. Arrange a group of substances in relative order of their entropies when given the formulas and states of the substances. (Prob. 14.28)
 *4. Use the relation $\Delta G = \Delta H - T\Delta S$ to calculate ΔH, T, or $\Delta S^{1\ atm}$

142 • UNIT 14–SPONTANEITY OF REACTION; ΔG AND ΔS

when given any two of the three quantities for a system at equilibrium (when $\Delta G = 0$). (Prob. 14.2)
5. Recognize that the values of $\Delta S^{1\ atm}$ and ΔH for reactions are essentially constant over reasonable temperature ranges.

14.4 The Gibbs-Helmholtz Equation
1. Recognize that spontaneous reactions, in general, tend toward greater disorder (increased entropy) and the evolution of energy ($\Delta H < 0$).
*2. Use the Gibbs-Helmholtz equation to calculate any variable, given the values of the other variables, or data to calculate one of the other variables. (Probs. 14.4, 14.33)

SUMMARY

There are three fundamental questions that a chemist wants answered when considering a particular reaction. They are: Will a reaction occur at a particular set of conditions? To what extent will the reaction occur? At what rate will it take place? This unit, along with Unit 4 (Thermochemistry), provides the basis for answering the first question. The principles that allow us to answer the questions regarding the extent and rate of a reaction will be discussed in Units 15 and 16.

1. Free Energy Change and Spontaneous Reactions

The energy changes associated with a chemical or physical process determine whether the process is capable of proceeding by itself once initiated. Such a process is known as a spontaneous process, and is capable of producing useful work. The fact that a process is capable of producing useful work does not mean that all, or for that matter any, of the work is actually harnessed to perform some function. Nor does it imply that a spontaneous process occurs rapidly. A spontaneous process can, if given the proper conditions and sufficient time, perform useful work.

The amount of useful work that can be obtained in a process occurring at constant temperature and pressure is limited by the difference in free energy between the products and the reactants, ΔG. For a reaction or process occurring at a constant pressure of 1 atm and a constant temperature, the reaction is spontaneous and capable of producing useful work when $\Delta G^{1\ atm}$ has a negative value. If $\Delta G^{1\ atm}$ is positive for such a process under these conditions, work must be done on the system and the reaction is said to be nonspontaneous; the reaction in the reverse direction is spontaneous. If $\Delta G^{1\ atm}$ is equal to zero, the reaction system is at equilibrium at 1 atm pressure.

Values of $\Delta G^{1\ atm}$ are related to the amounts of substances involved in a reaction and can be calculated from values of free energies of formation, $\Delta G_f^{1\ atm}$. Remember that the free energy of formation of an elementary substance at 1 atm and 25°C, like its ΔH_f, is zero.

Exercise 14.1

Consider the reaction in which ΔG at 25°C and 1 atm is -149.0 kcal:
$$N_2H_4(l) + 2O_2(g) \rightarrow N_2(g) + 2H_2O(l)$$

1. Using Table 14.2, calculate the free energy of formation of N_2H_4.
2. Comment on the stability of $N_2H_4(l)$, that is, predict whether N_2H_4 will decompose to N_2 and H_2 at 1 atm and 25°C.
3. Based on the above equation, calculate ΔG for the reaction of one gram of $N_2H_4(l)$ with excess oxygen at 1 atm and 25°C.
4. Would the value of $\Delta G^{1\ atm}$ for the above reaction be more or less negative when $H_2O(g)$ is formed rather than $H_2O(l)$ at these conditions.

Exercise 14.1 Answers

1. Remember that the free energy of formation of elementary substances is zero. Apply Equation 14.10 in the form:

$$\Delta G_f^{1\ atm}\ N_2H_4(l) = 2\ \Delta G_f^{1\ atm}\ H_2O(l) - \Delta G_{reaction}^{1\ atm}$$

$$= 2(-56.7\ \text{kcal}) - (-149.0\ \text{kcal}) = +35.6\ \text{kcal}$$

2. Since $\Delta G_f^{1\ atm}$ is +35.6 kcal, the reverse reaction, $N_2H_4(l) \rightarrow 2H_2(g) + N_2(g)$, must have a $\Delta G^{1\ atm}$ value of -35.6 kcal at these conditions. Since $\Delta G^{1\ atm}$ is negative, $N_2H_4(l)$ should decompose to the elements.

3. $\Delta G = -149.0$ kcal for one mole of N_2H_4 at these conditions.

   For one gram, $-149.0\ \dfrac{\text{kcal}}{\text{mole}} \times \dfrac{1\ \text{mole}}{32.0\ \text{g}} = -4.66\ \dfrac{\text{kcal}}{\text{g}}$

4. Referring to Table 14.2, $\Delta G_f^{1\ atm}\ H_2O(g) = -54.6$ kcal/mole. The value for $\Delta G_{reaction}^{1\ atm}$ would be less negative since this indicates that $H_2O(l) \rightarrow H_2O(g)$ is nonspontaneous at 25°C and 1 atm. $\Delta G^{1\ atm} = \Delta G_f^{1\ atm}\ H_2O(g) - \Delta G_f^{1\ atm}\ H_2O(l) = -54.6\ \text{kcal} - (-56.7\ \text{kcal}) = +2.1\ \text{kcal}$

2. Entropy, Entropy Change, and the Gibbs-Helmholtz Equation

For a given reaction, ΔG is dependent upon the temperature and the value of ΔH, as well as a state property called entropy. Entropy is a measure of the disorder or randomness of a system. The change in entropy, ΔS, is the difference between the entropies of the products and the reactants. A process tends to be favored when the system goes to a higher state of disorder; that is, when the entropy increases. On the other hand, a process is favored when there is a decrease in enthalpy, ΔH, the energy change associated with making and breaking bonds. The free energy change, ΔG, is the net result of these two effects. The Gibbs-Helmholtz equation, $\Delta G = \Delta H - T\Delta S$, relates these effects for a system at constant pressure, where T is the temperature in °K.

Since $\Delta G^{1\,atm}$ is equal to zero when a system is at equilibrium, $\Delta H = T\Delta S^{1\,atm}$. For phase changes, ΔH is the enthalpy change on fusion or on vaporization. $\Delta S^{1\,atm}$ for these physical processes can be calculated from this relation.

The effect of temperature, at constant pressure, on reaction spontaneity is dependent on the signs and magnitudes of both ΔH and ΔS. These effects are summarized in Table 14.5. The Gibbs-Helmholtz equation can be used to calculate $\Delta S^{1\,atm}$ when ΔH and $\Delta G^{1\,atm}$ are known or are calculated from ΔH_f and $\Delta G_f^{1\,atm}$ values at 25°C. Knowing $\Delta S^{1\,atm}$ and ΔH, $\Delta G^{1\,atm}$ can be calculated for temperatures other than 298°K. Another application of the Gibbs-Helmholtz equation is the determination of the temperature at which a reaction is at equilibrium; that is, when $\Delta G^{1\,atm}$ equals zero.

Exercise 14.2
1. Arrange the following in order of increasing entropy. $KBr(s)$, $Br_2(g)$, $K(s)$, $Br_2(l)$
2. Predict the sign of ΔS for each of the following:
   a. $2CO(g) + O_2(g) \rightarrow 2CO_2(g)$
   b. $Ag^+(aq) + Br^-(aq) \rightarrow AgBr(s)$
   c. $N_2O_5(s) \rightarrow 2NO_2(g) + \tfrac{1}{2}O_2(g)$
3. Estimate the normal boiling point of CCl_4. $\Delta H_{vap} = +7.75$. The molar entropies ($S^{1\,atm}$), in cal/°K, for liquid and gaseous CCl_4 are 51.3 and 74.1, respectively.
4. Consider the reaction:

   $C(s) + CO_2(g) \rightarrow 2CO(g)$

   a. Use Table 14.2 to calculate $\Delta G^{1\,atm}$ for the reaction at 25°C.
   b. Is the reaction spontaneous at these conditions? If not, estimate the minimum temperature at which this reaction becomes spontaneous. (You may need to refer to Table 4.1, Chapter 4, in the text.)

Exercise 14.2 Answers
1. There is greater probability for disorder in a KBr crystal than in the K crystal. This is due to the fact that two different kinds of ions are present in the KBr crystal. $Br_2(g)$ is certainly more disordered than the $Br_2(l)$. Both liquid and gaseous Br_2 are more disordered than either of the solids. The order should be

   $K(s) < KBr(s) < Br_2(l) < Br_2(g)$

2. Here we are comparing the entropies of the products to those of the reactants. When the disorder or randomness of the products is greater than that of the reactants, we predict an increase in entropy.
   a. The number of moles of gas decreases from three to two; predict a decrease in entropy.
   b. A solid is formed from two ions in solution. The entropy should decrease.
   c. One mole of solid produces 2 ½ moles of gas. Entropy should increase. (This is an example of a spontaneous endothermic reaction at 1 atm and 25°C, $\Delta H \approx 26$ kcal.)
3. The normal boiling point is the temperature at which equilibrium is established at 1 atm pressure for the process $CCl_4(l) \rightleftarrows CCl_4(g)$. Calculate $\Delta S^{1\ atm}$ for the process $CCl_4(l) \rightarrow CCl_4(g)$.

$$\Delta S^{1\ atm} = S^{1\ atm}(g) - S^{1\ atm}(l) = (74.1 - 51.3)\ cal/mole°K$$

$$= +22.8\ cal/mole°K.$$

Note that entropy units are in calories while ΔH is in kcal. The conversion 22.8 cal = 0.0228 kcal can be made. Since $G^{1\ atm} = 0$;

$$T = \frac{\Delta H_{vap}}{\Delta S^{1\ atm}} = \frac{+7.75\ kcal/mole}{+0.0228 kcal/mole°K} = 340°K = 67°C$$

4. a. $\Delta G^{1\ atm} = 2\Delta G_f^{1\ atm}\ CO(g) - \Delta G_f^{1\ atm}\ CO_2(g) = 2(-32.8\ kcal) - (-94.3\ kcal) = +28.7\ kcal$
   b. Since $\Delta G^{1\ atm}$ is positive, the reaction is not spontaneous. We can calculate the temperature at which the reaction is at equilibrium, $\Delta G = 0$, and interpret this temperature in terms of the temperature necessary for a spontaneous reaction. We will need to determine ΔH from ΔH_f values (Table 4.1) to calculate $\Delta S^{1\ atm}$ at 25°. Then we can calculate T when $\Delta G = 0$ as in Example 14.4.

$$\Delta H = 2\Delta H_f CO(g) - \Delta H_f CO_2(g) = 2(-26.4\ kcal) - (-94.1\ kcal)$$

$$= +41.3\ kcal$$

$\Delta G^{1\ atm}$ at 25°C = +28.7 kcal as calculated in (a).

$$\Delta S^{1\ atm} = \frac{+41.3\ kcal - 28.7\ kcal}{298°K} = +0.0423\ \frac{kcal}{°K}$$

At equilibrium $\Delta G^{1\ atm} = 0$;

$$T = \frac{\Delta H}{\Delta S^{1 \text{ atm}}} = \frac{+41.3 \text{ kcal}}{+0.0423 \frac{\text{kcal}}{°K}} = 976°K = 703°C$$

The temperature must be greater than 976°K for $\Delta G^{1 \text{ atm}}$ to have a negative value.

SUGGESTED ASSIGNMENT XIV

Text: Chapter 14

Problems: Numbers 14.1–14.4, 14.21, 14.23, 14.25, 14.27, 14.28, 14.30, 14.32, 14.33

Solutions to Assigned Problems Set XIV

14.1

a. To calculate ΔG, apply Equation 14.10.

$\Delta G^{1\ atm} = 2\Delta G_f^{1\ atm}\ SO_2(g) - 2\Delta G_f^{1\ atm}\ SO_3(g) = 2(-71.8\ kcal) - 2(-88.5\ kcal) = +33.4\ kcal$

b. Since ΔG is positive, the reaction is not spontaneous at these conditions.

c. For the reverse reaction, $\Delta G = -33.4\ kcal$

d. Note that in (a), $\Delta G^{1\ atm} = +33.4\ kcal/2\ moles\ SO_3$. Using conversion factors,

$$1.00g\ SO_3 \times \frac{1\ mole\ SO_3}{80.0g\ SO_3} \times \frac{33.4\ kcal}{2\ moles\ SO_3} = +0.209\ kcal$$

14.2

Liquid water is in equilibrium with its vapor at the boiling point of 100°C (373°K); ΔG = zero at equilibrium.

$\Delta H_{vap}/mole = 540\ cal/g \times 18.0g/mole = 9720\ cal/mole$

$$\Delta S_{vap}^{1\ atm} = \frac{\Delta H_{vap}}{T_b} = \frac{9720\ cal/mole}{373°K} = 26.1\ \frac{cal}{mole°K}$$

14.3

The sign of ΔS should be positive since there is an increase in the number of moles of gas (2 moles → 3 moles).

148 · UNIT 14–SPONTANEITY OF REACTION; ΔG AND ΔS

14.4

a. From molar heats of formation,

$\Delta H = 2\Delta H_f SO_2(g) - 2\Delta H_f SO_3(g) = 2(-71.0 \text{ kcal}) - 2(-94.5 \text{ kcal})$
$= +47.0 \text{ kcal}$

b. Using Equation 14.18,

$$\Delta S^{1 \text{ atm}} = \frac{\Delta H - \Delta G^{1 \text{ atm}}}{T} = \frac{+47.0 \text{ kcal} - (+33.4 \text{ kcal})}{298°K} = \frac{0.0456 \text{ kcal}}{°K}$$

$= 45.6 \text{ cal}/°K$

c. Assume that the values of ΔH and $\Delta S^{1 \text{ atm}}$ remain constant; the temperature is $500° + 273° = 773°K$

$\Delta G^{1 \text{ atm}} = \Delta H - T\Delta S^{1 \text{ atm}} = +47.0 \text{ kcal} - 773(0.0456 \text{ kcal}) = +11.8 \text{ kcal}$

The reaction is not spontaneous at this temperature, but note the temperature influence by comparing

$\Delta G^{1 \text{ atm}}$ at $298°K$ to $\Delta G^{1 \text{ atm}}$ at $773°K$.

d. The system is at equilibrium when $\Delta G^{1 \text{ atm}} = 0$ and

$$T = \frac{\Delta H}{\Delta S^{1 \text{ atm}}} = \frac{+47.0 \text{ kcal}}{+0.0456 \text{ kcal}/°K} = 1030°K$$

The temperature must be above $1030°K$ for the spontaneous decomposition of SO_3 since $\Delta G^{1 \text{ atm}}$ will have a negative value only when T exceeds $1030°K$.

Note that ΔH and ΔS must have the same units of energy in calculations using the Gibbs-Helmholtz equation.

14.21

a. The vaporization of water at 25°C is spontaneous when the pressure is less than 23.8 mmHg, is nonspontaneous when the pressure is greater than 23.8 mmHg, and is at equilibrium when the pressure is 23.8 mmHg, the equilibrium vapor pressure of water at 25°C.

b. Sodium chloride will dissolve spontaneously when the concentration is less than 6 M. The first reaction is spontaneous; the second is nonspontaneous.

14.23

a. The maximum useful work that could be obtained is 28 kcal.

b. The reaction could be reversed by doing 30 kcal of work but not 10 kcal.

14.25

We want to determine ΔG_f for the reaction

$$3Fe(s) + 2O_2(g) \rightarrow Fe_3O_4(s)$$

Applying Hess's Law, we multiply the first equation, and its value of ΔG by four, add it to the second equation, and simplify.

$$4[2Fe(s) + 1.5O_2(g) \rightarrow Fe_2O_3(s)]; \quad \Delta G = 4(-177 \text{ kcal})$$
$$\underline{4Fe_2O_3(s) + Fe(s) \rightarrow 3 Fe_3O_4(s); \quad \Delta G = -19 \text{ kcal}}$$
$$9Fe(s) + 6O_2(g) \rightarrow 3Fe_3O_4(s); \quad \Delta G = -727 \text{ kcal}$$

Divide the resulting equation (and ΔG) by 3 to obtain the equation and ΔG_f of Fe_3O_4.

$$3Fe(s) + 2O_2(g) \rightarrow Fe_3O_4(s); \Delta G = -727 \text{ kcal}/3 = -242 \text{ kcal/mole}$$

14.27

a. Note that this is the reverse reaction for the free energy of formation of $NH_3(g)$, except that it involves two moles of NH_3. The $\Delta G_f^{1 \text{ atm}}$ for one mole of $NH_3(g)$ is -4.0 kcal. For two moles decomposing, $\Delta G^{1 \text{ atm}} = 2 \text{ moles} \times \dfrac{+4.0 \text{ kcal}}{\text{mole}} = +8.0$ kcal. The reaction is not spontaneous.

b. Applying Equation 14.10,

$$\Delta G^{1 \text{ atm}} = 4\Delta G_f^{1 \text{ atm}} \, NO_2(g) + 6\Delta G_f^{1 \text{ atm}} \, H_2O(l) - 4\Delta G_f^{1 \text{ atm}}$$

$NH_3(g) = 4(+12.4 \text{ kcal}) + 6(-56.7 \text{ kcal}) - 4(-4.0 \text{ kcal}) = -274.6$ kcal

The reaction is spontaneous.

c. $\Delta G = 2\Delta G_f^{1\ atm}\ H_2O(l) - \Delta G_f^{1\ atm}\ SiO_2(s)$

$= 2(-56.7\ kcal) - (-192.4\ kcal) = +79.0\ kcal$

The reaction is not spontaneous.

14.28

To arrange these substances in order of increasing entropy, we need to consider the physical states as well as the number of atoms per formula unit for species in the same physical state. The solids will have less entropy than the gases and more complex species in the same physical state should have greater entropy.

$Li(s) < LiCl(s) < Ne(g) < Cl_2(g) < I_2(g)$

14.30

a. ΔS negative (gas molecules held on the surface)

b. ΔS positive (molecules or ions pass spontaneously from a more concentrated solution to a less concentrated solution)

c. ΔS negative (change from solid and gas to solid and liquid)

d. ΔS negative (pages becomes more ordered)

e. ΔS negative (gas to solid)

f. ΔS negative (form one large molecule from many smaller molecules)

14.32

a. Exothermic reactions are generally spontaneous; however, if ΔS is negative, an exothermic reaction may not be spontaneous at high temperatures.

b. If ΔS is positive, the $T\Delta S$ will make a negative contribution to ΔG; however, the sign of ΔH will also be important.

c. This is often true; however, the physical states must also be considered.

d. This statement is always true.

14.33

a. Use Table 4.1 to obtain ΔH_f values.

$\Delta H = \Delta H_f MgO(s) + \Delta H_f CO_2(g) - \Delta H_f MgCO_3(s)$

$= -143.8 \text{ kcal} + (-94.1 \text{ kcal}) - (-266 \text{ kcal}) = +28 \text{ kcal}$

b. $\Delta G^{1 \text{ atm}}$ at $298°K = \Delta G_f MgO(s) + \Delta G_f CO_2(g) - \Delta G_f MgCO_3(s)$

$= -136.1 \text{ kcal} + (-94.3 \text{ kcal}) - (-246 \text{ kcal}) = +16 \text{ kcal}$

c. At $298°K$, $\Delta S^{1 \text{ atm}} = \dfrac{\Delta H - \Delta G^{1 \text{ atm}}}{T} = \dfrac{28 \text{ kcal} - 16 \text{ kcal}}{298°K}$

$= 0.040 \text{ kcal}/°K$

d. Assume that ΔH and ΔS values remain constant.

$\Delta G^{1 \text{ atm}} = \Delta H - T\Delta S = 28 \text{ kcal} - (500°K)(0.040 \text{ kcal}/°K) = +8 \text{ kcal}$

The reaction is still nonspontaneous.

UNIT 15

Chemical Equilibrium in Gaseous Systems

The goals of this unit are to describe the characteristics of a system at equilibrium, investigate the relationships between the concentrations of all species in a system at equilibrium, predict the direction and extent of a reaction, and predict the effect of changes in conditions on the position of an equilibrium.

INSTRUCTIONAL OBJECTIVES

15.1 The N_2O_4–NO_2 Equilibrium System; Concept of K_c
  1. Recognize that the equilibrium constant, K_c, relates the equilibrium concentrations of the product and reactant species.
  *2. Distinguish between the number of moles of a species and the concentration of the species.
  3. Recognize that the square bracket notation [ ] is used to represent equilibrium concentrations in moles/liter.
  *4. Distinguish between initial concentrations and equilibrium concentrations; relate these concentrations to the chemical equation for a reaction.
  5. Realize that at a given temperature, the value of K_c is independent of the initial amounts of reactants and products, the direction from which equilibrium is approached, the volume of the container, and the total pressure.
  *6. Calculate the value of K_c for a reaction, given or having derived the equilibrium constant expression and knowing:
     a. the equilibrium concentrations of all species. (Prob. 15.28)
     b. the initial concentrations of all species (or information from

which the concentrations may be calculated) and the equilibrium concentration of one species. (Probs. 15.3, 15.29)

15.2 General Form and Properties of K_c
1. Realize that K_c has meaning only when associated with the chemical equation for the reaction.
*2. Write the equilibrium constant expression for K_c, given the equation for: a reaction in which all species are gases; a reaction in which one or more species are gases and one or more species are pure liquids or solids. (Prob. 15.1)
*3. Apply the Rule of Multiple Equilibria to obtain K_c for an overall reaction, given equations and values of K_c for other related reactions. (Probs. 15.2, 15.27)

15.3 Applications of K_c
*1. Predict the direction in which a reaction will proceed, given the equation for the reaction, the value of K_c, and initial numbers of moles and volume of the container, or initial concentrations. (Probs. 15.4, 15.31)
*2. Determine the extent to which a reaction occurs, given the equation for a reaction and value of K_c (or data from which K_c may be determined) by calculating:
   a. the equilibrium concentration of one species when given the equilibrium concentrations of all other species.
   b. the equilibrium concentrations of all species given their initial concentrations (or data to calculate these concentrations). (Probs. 15.5, 15.32)

15.4 Effect of Changes in Conditions Upon the Position of an Equilibrium
1. State LeChatelier's Principle.
*2. Given an equation for a reaction (including ΔH information and the phase of each species), qualitatively predict the direction of a shift in equilibrium when:
   a. the number of moles of reactant or product is changed.
   b. the volume is changed.
   c. the pressure is changed owing to a change in volume.
   d. the temperature is changed. (Prob. 15.6)
*3. Calculate the equilibrium concentrations of all species when given the value of K_c, initial concentrations, and the number of moles of reactants or products that are added to the original equilibrium system. (Prob. 15.32)
*4. Calculate the equilibrium concentrations of all species when given the value of K_c, the initial concentrations, and the change in volume or change in pressure from initial volume or pressure conditions. (Prob. 15.35)
*5. Apply Equation 15.12 to relate the value of K_c at a particular temperature, the value of ΔH, and the value of K_c at another temperature. (Prob. 15.7)

15.5 Relation Between the Free Energy Change and the Equilibrium Constant
1. Realize that for gaseous systems $\Delta G° = \Delta G^{1\ atm}$.
*2. Use Equation 15.13 to calculate the value of $\Delta G°$ when given a K value and the temperature. (Prob. 15.8)

SUMMARY

1. Chemical Equilibrium and Equilibrium Constants

The criterion for reaction spontaneity was discussed in Unit 14. You will recall that reactions occur spontaneously at a given temperature and pressure if ΔG is a negative quantity. When reactants are placed in a closed container and the reaction is initiated, the reactants produce products as long as the free energy is decreasing. As products form, the free energy change, ΔG, becomes less negative (i.e., as each additional mole of product forms, there is a smaller free energy change). Formation of products continues until the change in free energy becomes zero. At this point, the rate of the forward reaction is exactly equal to the rate of the reverse reaction. No net, observable change results, and a state of dynamic equilibrium is established.

For every equilibrium system at a given temperature, the product of the concentrations of the product species, raised to powers equal to the coefficients in the balanced chemical equation, divided by the concentrations of reacting species raised to the proper powers, equals a constant characteristic of the system. When the equilibrium concentrations are expressed in moles/liter, as indicated by [ ], this equilibrium constant is given the symbol K_c. The concentrations of pure solids and liquids are not included in the equilibrium expression. The numerical value of K_c for a particular equilibrium system is dependent only on the temperature. There are no restrictions on the individual equilibrium concentrations, the number of moles of substances introduced into the reaction container, nor on the direction from which the equilibrium state is approached. Remember, however, that equilibrium *concentrations* raised to the appropriate powers are used in the equilibrium expression, not the number of moles.

The Rule of Multiple Equilibria can be applied to obtain the value of K_c for a reaction from several related reactions and their K_c values. The values of K_c for each reaction are multiplied when the individual chemical equations are added to give the overall equation.

The establishment of equilibrium was described in terms of ΔG reaching a value equal to zero. The change in free energy is related to the equilibrium constant, K, by the expression $\Delta G° = -(2.3)RT \log_{10} K$, where $\Delta G°$ is the standard free energy change.

K_c = equilibrium constant

Exercise 15.1
1. Write the expression for K_c for
   a. $4NH_3(g) + 5O_2(g) \rightleftharpoons 4NO(g) + 6 H_2O(g)$
   b. $2CuO(s) + CO(g) \rightleftharpoons Cu_2O(s) + CO_2(g)$
2. Given that K_c for the reactions
   $2HI(g) \rightleftharpoons H_2(g) + I_2(g)$
   $2HF(g) \rightleftharpoons H_2(g) + F_2(g)$
   is 0.061 and 1.00×10^{-13}, respectively, calculate K_c for the reaction:

   $2HI(g) + F_2(g) \rightleftharpoons I_2(g) + 2HF(g)$

3. At 25°C, K for the reaction

   $CO(g) + Cl_2(g) \rightleftharpoons COCl_2(g)$

   is 6.1×10^{11}. Calculate $\Delta G°$.

Exercise 15.1   Answers

1. a. $K_c = \dfrac{[NO]^4 [H_2O]^6}{[NH_3]^4 [O_2]^5}$

   b. $K_c = \dfrac{[CO_2]}{[CO]}$

   The concentrations of the gaseous products appear in the numerator; the gaseous reactants in the denominator. The coefficients in the balanced equation become the exponents in the equilibrium expression. In (b), the pure solids Cu_2O and CuO do not appear in the expression. In (a), water is a gaseous product and appears in the equilibrium expression. If water is produced as a liquid, it would not appear in the expression.

2. An approach to this problem is to consider the reverse reaction involving HF. Writing the reaction in the reverse direction, we obtain HF on the product side and have F_2 on the reactant side.

   $H_2(g) + F_2(g) \rightleftharpoons 2HF(g); K_c = \dfrac{1}{1.00 \times 10^{-13}} = 1.00 \times 10^{13}$

   Adding the two equations and simplifying:

   $2HI(g) \rightleftharpoons H_2(g) + I_2(g);$ $K_c' = 0.061$
   $H_2(g) + F_2(g) \rightleftharpoons 2HF(g);$ $K_c'' = 1.00 \times 10^{13}$

   $2HI(g) + F_2(g) \rightleftharpoons I_2(g) + 2HF(g); K_c = K_c' K_c'' = (6.1 \times 10^{-2}) \times (1.00 \times 10^{13}) = 6.1 \times 10^{11}$

Note that: $(K_c')(K_c'') = \dfrac{[H_2][I_2]}{[HI]^2} \times \dfrac{[HF]^2}{[H_2][F_2]} = \dfrac{[I_2][HF]^2}{[HI]^2[F_2]}$

$= K_c$

3. Using the equation: $\Delta G° = -(2.30)RT \log_{10} K$, we have

$\Delta G° = -2.30 \times 1.99 \times 298 \times \log_{10} 6.1 \times 10^{11}$

$= -2.30 \times 1.99 \times 298 \times 11.79 = -16,100$ cal

2. Interpretation and Applications of K_c

The magnitude of an equilibrium constant indicates the extent to which a reaction proceeds toward completion. A large value of the equilibrium constant indicates that the reaction proceeds toward completion; a small value indicates that very little reaction occurs, and that the equilibrium system consists largely of reactants. Knowing the value of K_c and the original concentrations, it is possible to calculate the extent to which a reaction occurs. By comparing the concentration quotients for a given system to the value of K_c, we can determine if the system is at equilibrium: if not, we can find the direction the reaction will proceed to attain equilibrium.

LeChatelier's Principle can be used to qualitatively predict the effect of changes in conditions on a system that is at equilibrium. The principle states that if stress is applied to a system at equilibrium, the equilibrium shifts in such a way as to minimize the effect of the stress. If an equilibrium system is disturbed by adding a species (either reactant or products), the system will adjust in a direction to consume some of the added species. If a species is removed, the system will shift in the direction to produce that species. The extent to which a reaction occurs, as a result of adding or removing a species, can be calculated based on the value of K_c.

If the equilibrium system is disturbed owing to a decrease in the volume of the container, a reaction that results in a decrease in the total number of moles of gas will be favored. A reaction that results in an increase in the total number of moles of gas is favored when the volume of the container is increased. The quantitative effect of changes in the volume of a container, with the resulting changes in pressure, can again be related to the numerical value of the equilibrium constant.

The value of the equilibrium constant does not change because of pressure, volume, or concentration changes. The value of the constant does change, however, with a change in temperature. Qualitatively, an increase in temperature favors the endothermic process. Thus for an endothermic reaction, K_c will increase with an increase in temperature.

An increase in temperature results in a decrease in the value of K_c for an exothermic reaction. The magnitude of the temperature effect on the value of K may be calculated using Equation 15.12 in the text.

Exercise 15.2
1. Consider the reversible exothermic reaction

    $$4NH_3(g) + 5O_2(g) \rightleftharpoons 4NO(g) + 6H_2O(g), \Delta H = -217 \text{ kcal}$$

    If the system is at equilibrium, how will the following changes affect the number of moles of NO and the value of the equilibrium constant?
    a. More O_2 is added at constant temperature and pressure.
    b. More $H_2O(g)$ is injected into the container; the temperature and pressure remain constant.
    c. The volume of the container is decreased at constant temperature.
    d. The temperature is increased at constant volume.
2. Consider the equilibrium reaction

    $$CO(g) + H_2O(g) \rightleftharpoons CO_2(g) + H_2(g)$$

    At equilibrium at a given temperature, 0.80 mole CO_2, 0.80 mole H_2, 0.40 mole CO, and 0.40 mole H_2O are present in a one liter container. Calculate K_c.
3. Exactly 0.60 mole of each of the gases in Question 2 is added to a 2.0 liter container at the same temperature as in Question 2. Predict the direction in which the system will move to reach equilibrium at this temperature.
4. How many moles of CO_2 will be present when equilibrium is established in Question 3?
5. The reaction in Question 2 is studied at two different temperatures. The value of K_c at 700°C is 1.59; at 1000°C the value of K_c is found to be 0.60. Use Equation 15.12 to estimate the value of ΔH for this reaction.

Exercise 15.2 Answers
1. a. The reaction will occur that will consume O_2; more NO will be produced.
   b. The reaction that consumes H_2O will occur; the number of moles of NO will decrease.
   c. The reaction that results in a decrease in the total number of moles will occur (from 10 to 9 in this case). The number of moles of NO will decrease.
   d. An increase in temperature favors the endothermic reaction. The endothermic reaction is the formation of NH_3 and O_2. The number of moles of NO will decrease.

The only change that affects K_c is the increase in temperature. Since this is an exothermic reaction, K_c will decrease

2. $K_c = \dfrac{[CO_2][H_2]}{[CO][H_2O]} = \dfrac{(0.80)(0.80)}{(0.40)(0.40)} = 4.0$

3. First calculate the initial concentrations, then compare the original concentration quotient to K_c, 4.0, to predict in which direction the reaction will proceed.

Initial concentrations of all species = 0.60 mole/2.0 liters = 0.30 mole/liter

$$\dfrac{(\text{orig conc } CO_2)(\text{orig conc } H_2)}{(\text{orig conc } CO)(\text{orig conc } H_2O)} = \dfrac{(0.30)(0.30)}{(0.30)(0.30)} = 1.0$$

Since the quotient is smaller than K_c, 4.0, the reaction will proceed to produce CO_2 and H_2.

4. Let x = number of additional moles/liter of CO_2 that will form. This must also produce an additional x moles/liter of H_2. Thus $[CO_2] = [H_2] = 0.30 + x$

In forming x moles/liter of CO_2, x moles/liter of CO and x moles/liter of H_2O are consumed. $[CO] = [H_2O] = 0.30 - x$

$$K_c = \dfrac{[CO_2][H_2]}{[CO][H_2O]} = 4.0 = \dfrac{(0.30 + x)^2}{(0.30 - x)^2}$$

Since this is a perfect square, take the square root of both sides of the equation and solve for x.

$(4.0)^{½} = \dfrac{0.30 + x}{0.30 - x} = 2.0;\ 0.30 + x = 2.0(0.30 - x) = 0.60 - 2.0x;$

$3.0x = 0.30;\ x = 0.10$

The concentration of CO_2 at equilibrium = (0.30 + 0.10) moles/liter.

The number of moles of CO_2 present at equilibrium in the 2.0 liter container is 0.40 mole/liter × 2.0 liters = 0.80 mole

5. Let: $T_1 = (700 + 273)°K = 973°K;\ K_1 = 1.59$

$T_2 = (1000 + 273)°K = 1273°K;\ K_2 = 0.60$

$$\log_{10} \dfrac{K_2}{K_1} = \dfrac{\Delta H}{(2.30)(1.99)} \left[\dfrac{T_2 - T_1}{T_2 \times T_1}\right]$$

$$\log_{10}\left(\dfrac{0.60}{1.59}\right) = \dfrac{\Delta H}{(2.30)(1.99)} \left[\dfrac{1293 - 973}{1293 \times 973}\right]$$

$$\log_{10} 3.77 \times 10^{-1} = \frac{\Delta H}{4.58} (2.4 \times 10^{-4})$$

$$\log_{10} 3.77 - 1 = 0.58 - 1 = -0.42 = \Delta H(5.24 \times 10^{-5})$$

$$\Delta H = \frac{-0.42}{5.24 \times 10^{-5}} = -8.0 \times 10^3 \text{ cal.}$$

SUGGESTED ASSIGNMENT XV

Text: Chapter 15

Problems: Numbers 15.1–15.8, 15.27–15.29, 15.31, 15.32, 15.35, 15.39

Solutions to Assigned Problems Set XV

15.1

When writing expressions for K_c, the concentrations of the products are placed in the numerator and the concentrations of the reactants in the denominator. The exponents of the concentration terms are the coefficients in the balanced equation. No term is included for the solids.

a. $K_c = \dfrac{[CO]^2 [O_2]}{[CO_2]^2}$

b. $K_c = \dfrac{[CO_2]^2}{[CO]^2}$

15.2

a. $K_a K_b = \dfrac{[O_2] \times [CO]^2}{[CO_2]^2} \times \dfrac{[CO_2]^2}{[CO]^2} = [O_2] = K_c$

Reaction (a) + Reaction (b) = Desired Reaction

15.3

We need to consider the coefficients in the balanced equation to determine the number of moles of products formed and the number of moles of HI that decompose. If the initial concentration of HI is 0.50 M and 0.10 M is the equilibrium concentration, 0.40 mole/ℓ must decompose. Using the coefficients in the balanced equation as conversion factors,

$$0.40 \text{ mole/ℓ HI} \times \dfrac{1 \text{ mole/ℓ } H_2}{2 \text{ mole/ℓ HI}} = 0.20 \text{ mole/ℓ } H_2$$

The concentration of $[H_2] = [I_2] = 0.20$ M

	Initial Conc (mole/liter)	Change in Conc (mole/liter)	Equil Conc (mole/liter)
HI	0.50	−0.40	0.10
H_2	0.00	+0.20	0.20
I_2	0.00	+0.20	0.20

$$K_c = \frac{[H_2][I_2]}{[HI]^2} = \frac{(0.20)(0.20)}{(0.10)^2} = 4.0$$

15.4

For these problems, we compare the original concentration quotient to K_c, 4.0, to decide in which direction the reaction will proceed.

a. $\dfrac{(\text{orig conc } H_2)(\text{orig conc } I_2)}{(\text{orig conc HI})^2} = \dfrac{(0.00)^2}{(0.25)^2} = 0 < K_c$

The reaction will proceed to the right to produce H_2 and I_2.

b. orig conc quotient $= \dfrac{(0.50)(0.20)}{(0.10)^2} = 10$

Since the original concentration quotient, 10, is larger than K_c, 4, the reverse reaction must take place. The H_2 and I_2 will react to produce additional HI until the quotient decreases to a value of 4.0.

15.5

For the system: $2HI(g) \rightleftharpoons H_2(g) + I_2(g)$

$$K_c = \frac{[H_2][I_2]}{[HI]^2} = 4.0$$

Let $X = [H_2]$; X must also equal the equilibrium concentration of I_2 for this problem. To produce X moles of H_2 per liter, 2X moles of HI must decompose. Summarizing these relationships in a table similar to Table 15.1, we have:

	Initial Conc (moles/liter)	Change in Conc (moles/liter)	Equil Conc (moles/liter)
HI	0.25	$-2X$	$0.25 - 2X$
H_2	0.00	$+X$	X
I_2	0.00	$+X$	X

$$K_c = 4.0 = \frac{(X)(X)}{(0.25 - 2X)^2}$$

Note that this is a perfect square. Taking the square root of both sides, we have:

$$(4.0)^{\frac{1}{2}} = 2.0 = \frac{X}{0.25 - 2X}$$

Solving for X: $X = 2(0.25 - 2X) = 0.50 - 4X$; $5X = .50$; $X = 0.10$. The equilibrium concentration of $H_2 = 0.10$ M.

15.6

We apply LeChatelier's Principle to qualitatively predict the direction in which the equilibrium will shift when changes are made on the following equilibrium system.

$2SO_2(g) + O_2(g) \rightleftharpoons 2SO_3(g); \Delta H = -47.0$ kcal

a. If SO_3 is added, the reaction that consumes SO_3 will occur. The reverse reaction will occur, so the system will shift towards the left.

b. Removing SO_3 will cause the equilibrium to shift to produce SO_3. The reaction will proceed to the right.

c. When the volume is increased, the reaction that increases the number of moles will occur. The production of three moles of gas ($2SO_2$ and $1O_2$) from two moles will occur. The reaction will shift to the left.

d. An increase in temperature always favors the endothermic process. Since the forward reaction is exothermic (ΔH is negative), increasing the temperature causes the equilibrium to shift to absorb energy, or to the left in this case. Note that a change in temperature is the only factor that changes the value of K_c. An

164 • UNIT 15–CHEMICAL EQUILIBRIUM IN GASEOUS SYSTEMS

increase in temperature will lead to a decrease in K_c for this system.

e. Since N_2 does not react, a pressure increase due to the addition of N_2 does not affect the equilibrium. Only pressure changes due to a change in volume will affect this equilibrium.

15.7

Equation 15.12 relates K_c at two different temperatures. Note that ΔH must be in calories (-47.0 kcal $= -47,000$ cal) when $R = 1.99$ cal/mole°K. Taking T_2 as the higher temperature:

$$\log_{10} \frac{K_2}{K_1} = \frac{\Delta H}{(2.30)(1.99)} \left[\frac{T_2 - T_1}{T_2 \times T_1} \right] = \frac{-47,000}{(2.30)(1.99)} \times \left[\frac{1200 - 1000}{1200 \times 1000} \right] = -1.71$$

Taking the antilog of -1.71, $K_2/K_1 = 1.95 \times 10^{-2}$; $K_2 = (1.95 \times 10^{-2})(100) = 1.95$

15.8

Equation 15.13 relates $\Delta G°$ to K. Substituting, we have:

$\Delta G° = -(2.30)(1.99)\, T \log_{10} K = -(2.3)(1.99)(1000)(\log_{10} 100) = -9150$ cal $= -9.15$ kcal

15.27

Using the rule of multiple equilibria, we write:

$H_2(g) + S(s) \rightleftharpoons H_2S(g)$; $\qquad K_1 = 1.0 \times 10^{-3}$

$SO_2(g) \rightleftharpoons S(s) + O_2(g)$; $\qquad K_2 = \dfrac{1}{5.0 \times 10^6}$

$H_2(g) + SO_2(g) \rightleftharpoons H_2S(g) + O_2(g)$;

$K_c = K_1 \times K_2 = (1.0 \times 10^{-3}) \times \left(\dfrac{1}{5.0 \times 10^6} \right) = 2.0 \times 10^{-10}$

15.28

First calculate the equilibrium concentrations in moles/liter.

$[SO_2] = 0.40$ mole/2.0 liters $= 0.20$

$[O_2] = 0.030$ mole/2.0 liters $= 0.015$

$[SO_3] = 1.00$ mole/2.0 liters $= 0.50$

Substituting into the equilibrium expression:

$$K_c = \frac{[SO_3]^2}{[SO_2]^2[O_2]} = \frac{(0.50)^2}{(0.20)^2(0.015)} = 420$$

15.29

The number of moles per liter of SO_3 that react is $(1.20 - 0.42)$ moles $= 0.78$ mole. The number of moles per liter of O_2 formed is 0.78 moles $SO_3 \times \frac{1 \text{ mole } O_2}{2 \text{ moles } SO_3} = 0.39$ mole. Since one mole of SO_2 is produced for every mole of SO_3 that decomposes, 0.78 mole per liter is present at equilibrium. Summarizing in a table similar to Table 15.1, we have:

	Initial Conc (moles/liter)	Change in Conc (moles/liter)	Equil Conc (moles/liter)
SO_3	1.20	-0.78	0.42
O_2	0	+0.39	0.39
SO_2	0	+0.78	0.78

Substituting into the equilibrium expression:

$$K_c = \frac{[SO_3]^2}{[SO_2]^2[O_2]} = \frac{(0.42)^2}{(0.78)^2(0.39)} = 0.74$$

15.31

a. We compare the initial concentration quotient to K_c, 1.6, to predict the direction the system will move to reach equilibrium.

$$\frac{(\text{orig conc } H_2O)^2 (\text{orig conc } Cl_2)^2}{(\text{orig conc } HCl)^4 (\text{orig conc } O_2)} = \frac{(0.20)^2 (0.20)^2}{(0.20)^4 (0.20)} = 5.0$$

Since $5.0 > 1.6$, the reaction will proceed to the left.

b. The approach is similar to (a) except that we first need to calculate the concentrations.

HCl: 1.20 moles/4.0 liters = 0.30 M

O_2: 0.60 mole/4.0 liters = 0.15 M

H_2O: 1.40 moles/4.0 liters = 0.35 M

Cl_2: 0.80 mole/4.0 liters = 0.20 M

$$\frac{(\text{orig conc } H_2O)^2 (\text{orig conc } Cl_2)^2}{(\text{orig conc } HCl)^4 (\text{orig conc } O_2)} = \frac{(0.35)^2 (0.20)^2}{(0.30)^4 (0.15)} = 4.0$$

Since $4.0 > 1.6$, the reaction will proceed to the left to establish equilibrium.

15.32

a. Let $x = [SO_3] = [NO]$. Note that brackets indicate equilibrium concentrations in moles/liter.

Since x moles of SO_2 and x moles of NO_2 must react to produce x moles of SO_3 and x moles NO, $[SO_2] = [NO_2] = (3.0 \times 10^{-3} - x)$

$$K_c = \frac{[SO_3][NO]}{[SO_2][NO_2]} = \frac{x^2}{(3.0 \times 10^{-3} - x)^2} = 9.0$$

Since this is a perfect square, take the square root of both sides of the equation to give

$$\frac{x}{(3.0 \times 10^{-3} - x)} = 3.0$$

Solving for x: $x = 3.0 (3.0 \times 10^{-3} - x) = 9.0 \times 10^{-3} - 3.0 x$

$4x = 9.0 \times 10^{-3}$; $x = 2.25 \times 10^{-3}$

$[SO_3] = [NO] = 2.2 \times 10^{-3}$

$[SO_2] = [NO_2] = (3.0 \times 10^{-3}) - (2.25 \times 10^{-3}) = 0.75 \times 10^{-3}$
$= 7.5 \times 10^{-4}$

b. $K_c = \dfrac{[SO_3][NO]}{[SO_2][NO_2]} = 9.0$. Since the initial concentrations of all species are equal, 3.0×10^{-3} M, the reaction must proceed to the right to produce SO_3 and NO in order to equal the value of the K_c at equilibrium.

Let x = number of moles of SO_2 and NO_2 that react. The equilibrium concentrations of SO_2 and NO_2 are

$[SO_2] = [NO_2] = 3.0 \times 10^{-3} - x$

$[SO_3] = [NO] = 3.0 \times 10^{-3} + x$

$K_c = \dfrac{(3.0 \times 10^{-3} + x)^2}{(3.0 \times 10^{-3} - x)^2} = 9.0$

Taking the square root of both sides of the equation.

$\dfrac{3.0 \times 10^{-3} + x}{3.0 \times 10^{-3} - x} = 3.0$

Solving for x: $3.0 \times 10^{-3} + x = 3.0(3.0 \times 10^{-3} - x) = 9.0 \times 10^{-3} - 3x$; $4x = 6.0 \times 10^{-3}$; $x = 1.5 \times 10^{-3}$

$[SO_2] = [NO_2] = 3.0 \times 10^{-3} - 1.5 \times 10^{-3} = 1.5 \times 10^{-3}$

$[SO_3] = [NO] = 3.0 \times 10^{-3} + 1.5 \times 10^{-3} = 4.5 \times 10^{-3}$

15.35

For the reaction $N_2O_4(g) \rightleftharpoons 2NO_2(g)$

$K_c = \dfrac{[NO_2]^2}{[N_2O_4]} = 0.36$

Starting with one mole of N_2O_4 in a one liter container, let x = number of moles of N_2O_4 that react. Then 2x = number of moles of NO_2 formed. The remaining number of moles of $N_2O_4 = 1.0 - x$. Since the container has a volume of 1.0 liter, $[NO_2] = \dfrac{2x \text{ moles}}{1.0 \text{ liter}} = 2x$;

$[N_2O_4] = \dfrac{(1.0 - x) \text{ moles}}{1.0 \text{ liter}} = 1.0 - x$

Substituting into the equlibrium expression:

$$\frac{[NO_2]^2}{[N_2O_4]} = 0.36 = \frac{(2x)^2}{(1.0-x)}; \; 4x^2 = 0.36(1.0-x) = 0.36 - 0.36x \text{ or}$$

$$4x^2 + 0.36x - 0.36 = 0$$

Substituting into the formula $x = \dfrac{-b \pm \sqrt{b^2 - 4ac}}{2a}$ where $a = 4$; $b = 0.36$; $c = -0.36$

$$x = \frac{-0.36 \pm \sqrt{(0.36)^2 - 4(4)(-0.36)}}{2(4)} = \frac{-0.36 \pm \sqrt{5.89}}{8}$$

$$= \frac{-0.36 \pm 2.43}{8} = \frac{2.07}{8} \text{ or } \frac{-2.79}{8} \text{ (i.e., 0.26 or } -0.35).$$

A value of -0.35 for x is not possible. Therefore, $x = 0.26$

$1.0 - 0.26 = 0.74 = [N_2O_4]$; $[NO_2] = 2(0.26) = 0.52$

These are the values tabulated in Table 15.2.

15.39

Writing Equation 15.13 in the following form and substituting the values for $\Delta G°$, in calories, and T, we have:

$$\log_{10} K = \frac{-\Delta G°}{(2.30)(1.99)T} = \frac{-10,000}{(2.30)(1.99)(5.00)} = -4.37$$

Taking antilogs, $\log_{10} K = 0.63 - 5$; $K = 4.2 \times 10^{-5}$

For $\Delta G° = -10$ kcal $= -10,000$ cal:

$$\log_{10} K = \frac{-(-10,000)}{(2.30)(1.99)(500)} = 4.37$$

Taking antilogs, $\log_{10} K = 0.37 + 4$; $K = 2.3 \times 10^4$

UNIT 16

Rates of Reaction

The goal of this unit is to consider the factors (concentration, temperature, catalyst) that affect reaction rates, and develop a model, on the molecular level, of the mechanism by which a reaction occurs.

INSTRUCTIONAL OBJECTIVES

16.1 Meaning of Reaction Rate
1. Realize that there is no direct correlation between the rate of a reaction and the free energy change or the magnitude of the equilibrium constant.
2. Explain the meaning of reaction rate, average rate, initial rate, initial concentration.
*3. Express the average rate of reaction over a specific time period or the rate at a particular time, given the concentration of a species at various times. (Prob. 16.1)

16.2 Dependence of Reaction Rate Upon Concentration
*1. Given data on initial rates of reaction as a function of concentration, and the equation for the reaction:
 a. derive the rate expression. (Prob. 16.23)
 b. determine the order of reaction with respect to each reactant and the overall order of reaction. (Prob. 16.23a)
 c. calculate the specific rate constant. (Prob. 16.23b, 16.24)
*2. Calculate the rate at which a product is formed, or a reactant is consumed, given the rate expression and concentration data. (Probs. 16.2, 16.22)
3. Realize that the rate expression is obtained from experimental data, not from the coefficients in a balanced equation.
*4. For a first-order reaction, use Equation 16.5 to calculate:
 a. the concentration of a reactant remaining after time t, when

given the rate constant and initial concentration. (Prob. 16.27c)
b. the time required for the concentration to decrease to a given value, when given the rate constant and the initial concentration. (Prob. 16.3)
*5. For a first order reaction, use Equation 16.6 to determine the half-life or the rate constant, given one of the two quantities. (Probs. 16.27, 16.28)
*6. Graphically determine the order of a reaction, given concentrations as a function of time. (Prob. 16.26)

16.3 Dependence of Reaction Rate Upon Temperature
  1. Define the terms activation energy and activated complex.
 *2. Use the Arrhenius equation (Equation 16.10) to calculate:
    a. the activation energy when given rate constants at two different temperatures. (Prob. 16.4)
    b. the rate constant at one temperature, given the value of the rate constant at another temperature, and the activation energy.
    c. the temperature at which k will have a certain value when given the values for E_a, k_1, and T_1.
 *3. Sketch, label, and interpret a reaction profile diagram similar to Figure 16.6, given the activation energy and the enthalpy change. (Prob. 16.32)

16.4 Catalysis
 *1. Interpret, on a molecular basis, a reaction profile diagram similar to Figure 16.7 for a catalyzed and uncatalyzed reaction.
  2. Recognize that a catalyst does not affect the enthalpy or free energy changes for a reaction.
  3. Briefly describe the action of an enzyme in catalyzing biochemical reactions and the action of inhibitors that decrease enzyme activity.

16.5 Collision Theory of Reaction Rates
  1. For a bimolecular process, relate the total kinetic energy of collision and the orientation of the colliding molecules to the activation energy for a reaction.
  2. Relate the temperature dependence of reaction rate to the distribution of molecular energies, Figure 5.13.
 *3. Evaluate the factor Z in Equation 16.14 in terms of concentration of reactants in accounting for the effect of concentration on the reaction rate.

16.6 Reaction Mechanisms
  1. Recognize that many reactions occur by way of multistep mechanisms.

2. Define the following terms: rate determining step, chain reaction, chain initiation, chain termination, chain propagation.
3. Describe the factors affecting the rate of a surface reaction.
*4. Determine whether or not a particular mechanism for a reaction is consistent with the rate expression. (Prob. 16.5)
*5. Determine the rate expression, given a reaction mechanism. (Prob. 16.35)

SUMMARY

1. Reaction Rates and Rate Expressions

Three fundamental questions concerning chemical reactions were posed in the Unit 14 Summary. The first question was concerned with the spontaneity of reaction. This was discussed in terms of the change in free energy. The second question dealt with the extent of a reaction. This was described by the equilibrium of the system. Even though a reaction may be thermodynamically favorable, that is, ΔG is negative and K has a large value, significant amounts of products will not be produced unless the reaction proceeds at a reasonable rate.

An ordinary chemical equation indicates the nature of the reactants and the products, but not how these substances are consumed or produced. To understand how a reaction proceeds and, thus, the factors that influence the rate of a reaction, we need to consider the behavior of matter on the molecular level, and the pathway or mechanism by which the reaction occurs.

The reaction rate is defined as a change in concentration of a reactant or product with time. In general, the rate of reaction is found to be dependent upon the nature of the reactants, the concentration of the reactants, the temperature, and the presence of a catalyst. The rate of reaction is always a positive quantity that has a unit of concentration per time, such as moles liter^{-1} min^{-1} or moles liter^{-1} second^{-1}. The rate of reaction depends upon the time at which the rate is measured and the time interval over which the measurement is made, since the rate of reaction is continuously changing from the initial rate to the rate at completion or at equilibrium.

Exercise 16.1

The rate of decomposition of nitrogen dioxide, NO_2, is investigated at a certain temperature. At 2.0 minutes after the study begins, the concentration of NO is found to be 0.30 M. At 4.0 minutes after the study begins, the concentration is 0.70 M. The equation for the reaction is:

$$2NO_2(g) \rightarrow 2NO(g) + O_2(g)$$

1. Determine the average rate of formation of NO.
2. Determine the average rate of decomposition of NO_2.
3. Determine the average rate of formation of O_2.

Exercise 16.1 Answers
1. The average rate of formation of NO over this time period is:

$$\text{average rate} = \frac{\Delta \text{ conc NO}}{\Delta t} = \frac{0.70 \text{ M} - 0.30 \text{ M}}{4.0 \text{ min} - 2.0 \text{ min}} = \frac{0.40 \text{ M}}{2.0 \text{ min}} = 0.20$$

mole liter^{-1} min^{-1}

2. Since there is a one to one mole relationship between NO_2 and NO, 0.40 M of NO_2 must decompose to form 0.40 M NO.

$$\text{average rate} = \frac{\Delta \text{ conc } NO_2}{\Delta t} = \frac{\Delta \text{ conc NO}}{\Delta t}$$

Note that rate of reaction is always a positive number. Therefore, when the rate is expressed in terms of a reactant (in which the conc NO_2 after time t is less than the conc NO_2 at an earlier time), the change in concentration of the reactant is negative.

$$-\frac{\Delta \text{ conc } NO_2}{\Delta t} = -\frac{(-0.40 \text{ M})}{2.0 \text{ min}} = 0.20 \text{ mole liter}^{-1} \text{ min}^{-1}$$

3. For every two moles of NO produced, one mole of O_2 forms. Thus the rate at which NO forms is twice the rate at which O_2 forms.

$$\frac{\Delta \text{ conc NO}}{\Delta t} = 2\left(\frac{\Delta \text{ conc } O_2}{\Delta t}\right); 0.20 \text{ mole liter}^{-1} \text{ min}^{-1}$$

$$= 2\left(\frac{\Delta \text{ conc } O_2}{\Delta t}\right); \frac{\Delta \text{ conc } O_2}{\Delta t} = \frac{0.20 \text{ mole liter}^{-1} \text{ min}^{-1}}{2}$$

$$= 0.10 \text{ mole liter}^{-1} \text{ min}^{-1}$$

The rate of a reaction is often proportional to the concentrations of the reactants raised to some power. The proportionality constant k, the rate constant, is dependent on the temperature, and has units that depend on the concentration units and the powers to which these concentration units are raised. The power to which the concentration of each reactant is raised in the rate expression is called the order of reaction with respect to that reactant. The overall order of reaction is the sum of the exponents.

The order of a reaction cannot be obtained directly from the coefficients in the balanced equation. The actual values of the exponents must be determined experimentally. When a reaction involves several reactants, experiments are usually designed in which a series of reactions is studied. The initial concentration of one of the reactants is changed and the effect on the rate of each reaction is observed. This approach is illustrated in Example 16.1 in the text, in which one reactant is involved in a decomposition reaction. Exercise 16.2 illustrates the same principles for a reaction involving several reactants. Notice that once the rate expression has been determined, k can be evaluated from the rate of reaction at any set of concentrations. In addition, the rate can be estimated for any set of concentrations if the value of k is known.

Exercise 16.2

For the reaction: $A(g) + B(g) \rightarrow C(g)$, the initial rates of formation of C (mole liter^{-1} min^{-1}) at various concentrations (mole liter^{-1}) are:

Experiment	conc A	conc B	rate
1	0.10	0.10	0.10
2	0.10	0.20	0.40
3	0.20	0.20	0.80
4	0.20	0.40	3.20

1. Determine the order of reaction with respect to A; with respect to B. What is the overall order of reaction?
2. Write the rate expression for this reaction.
3. Calculate the value of the rate constant.
4. What will be the initial rate of formation of C if the initial concentration of A is 0.40 M and that of B is 0.60 M?

Exercise 16.2   Answers

1. The approach here is to compare either experiments 1 and 2 or experiments 3 and 4 to determine the effect of the concentration of B on the rate. In these two sets of experiments, the concentration of reactant A is held constant and the concentration of B is changed. When the concentration of B is changed from 0.10 M to 0.20 M, the rate of formation of C changes by a factor of four. Comparing experiment 3 with experiment 4, a doubling of the concentration of B (with the concentration of A held constant) also results in a four-fold increase in the rate of formation of C. The rate is second order with respect to B.

   In experiments 2 and 3, the concentration of B is held constant, and the concentration of A is doubled. The rate also changes by a factor of two. The reaction is first order with respect to A. The overall order is 1 + 2 = 3.

2. The rate expression is:

$$\text{rate} = \frac{\Delta \text{conc C}}{\Delta t} = k \text{(conc A)(conc B)}^2$$

3. To calculate the value of k, substitute the data for one of the experiments into the rate expression. Using values from experiment 1:

$$k = \frac{\text{rate}}{\text{(conc A)(conc B)}^2} = \frac{0.10 \text{ mole liter}^{-1} \text{ min}^{-1}}{(0.10 \text{ mole liter}^{-1})(0.10 \text{ mole liter}^{-1})^2}$$

$$= 1.0 \times 10^2 \frac{\text{liter}^2}{\text{mole}^2 \text{ min}}$$

4. Substitute into the rate expression and use the value of k calculated above.

$$\text{rate} = k \text{(conc A)(conc B)}^2 = 1.0 \times 10^2 \frac{\text{liter}^2}{\text{mole}^2 \text{ min}} \times$$

$$\left(4.0 \times 10^{-1} \frac{\text{mole}}{\text{liter}}\right) \times \left(6.0 \times 10^{-1} \frac{\text{mole}}{\text{liter}}\right)^2 =$$

$$1.4 \times 10^1 \frac{\text{mole}}{\text{liter min}}$$

The rate expression indicates how the rate of a reaction changes with a change in concentration, but in this form it does not relate the actual concentrations of reactants or products at a particular time during the course of a reaction. When the rate depends on only one reactant, concentration-time data are treated graphically to determine order of reaction. For the reaction A → products, characteristics of zero, first, and second order reactions are summarized as follows:

Order	Rate Expression	Linear Plot	Conc-time Relation
0	rate = k	X vs. t	$X_0 - X = kt$
1	rate = k X	$\log_{10} X$ vs. t	$\log_{10} \frac{X_0}{X} = \frac{kt}{2.30}$
2	rate = k X^2	$\frac{1}{X}$ vs. t	$\frac{1}{X} - \frac{1}{X_0} = kt$

(X and X_0 = conc reactant at time t and time t = 0, respectively)

The concept of half-life, the time required to reduce the concentration to one-half the original value (conc X_t = ½ conc X_0), is an important characteristic. For a first order reaction, $t_{½} = \dfrac{0.693}{k}$, the half-life is independent of the concentration. Examples 16.2 and 16.3 in the text illustrate the treatment of concentration-time data to determine order of reaction and calculations involving a first order reaction.

2. Temperature, Catalysts, and Collision Theory

One approach to the explanation of reaction rates is to consider that reactions occur as a result of collisions between reacting species. Collision theory proposes that the rate of reaction is proportional to the number of collisions that occur per second in a unit volume of reaction mixture. Although the number of collisions may be extremely large, only those collisions that equal or exceed a particular kinetic energy, and have the proper molecular orientation, are effective in producing a reaction. A plot of potential energy vs. path of reaction, as in Figure 16.6, indicates that the potential energy increases as the reacting species approach and collide. This increase in potential energy arises from the conversion of the kinetic energies of the colliding molecules. A reaction occurs when atoms in the molecules are rearranged, as some bonds are broken and other bonds formed. The minimum potential energy required to bring the reactants to the point where they can rearrange to form product molecules is called the activation energy. The unstable, transient species formed at the peak of the energy curve is called the activated complex. Note that the relationship between the enthalpy change, ΔH, for the overall reaction, and the activation energy for the forward reaction, E_a, and the reverse reaction, E_a', is: $\Delta H = E_a - E_a'$.

The activation energy depends on the conversion of the kinetic energy of the colliding molecules into potential energy. The distribution of molecular speeds (and thus kinetic energies) among a sample of gas molecules at two different temperatures is indicated in Figure 5.13. Notice that the relative numbers of molecules of higher energy increase at a higher temperature. The temperature dependence of reaction rate is related to the distribution of molecular energies and is expressed in terms of the activation energy and reaction rate by Equation 16.9 and Equation 16.16. Equation 16.9 may be used directly to obtain the activation energy from a plot of $\log_{10} k$ vs. $1/T$ or used in the form of Equation 16.10. This relationship is illustrated in Example 16.4.

A catalyst is a substance that increases the reaction rate without being permanently changed itself. The catalyst functions to decrease the activation energy and to provide an alternate pathway for the reaction. Catalysts accelerate both forward and reverse reactions by lowering the activation energy for each reaction by the same amount. A catalyst does not alter the value of K_c or ΔG for a reaction. Enzymes are examples of catalysts that promote specific biochemical reactions. The action of an enzyme may be diminished by substances on its surface known as inhibitors. Certain substances that exist in a separate phase from that of the reactants and products can behave as catalysts. The surface of these substances adsorbs reactant species and provides a mechanism that lowers the activation energy for the reaction.

3. Reaction Mechanisms

A reaction mechanism is a description of the path or a series of steps by which a reaction occurs. The mechanism is represented by a series of equations for the individual reactions that occur. The slowest reaction in the sequence is called the rate determining step because the overall rate of reaction can not be any greater than that of the slowest step.

In writing a mechanism, a series of steps is proposed which add together to give the overall reaction. The predicted dependence upon reactant concentrations must be consistent with the experimentally determined rate expression. The rate equation for a step in a reaction has the order for each reactant equal to the coefficient of that reactant in that step. Most steps in a reaction involve a collision between only two molecules, frequently producing reactive intermediate species. Intermediate species that are present as reactants in the rate determining step can not appear in the final rate equation. These species can often be eliminated by the use of an equilibrium condition.

Example 16.5 and the following exercise illustrate the evaluation of proposed mechanisms to determine if they are consistent with the rate laws.

Exercise 16.3

For the reaction: $2NO(g) + O_2(g) \rightarrow 2NO_2(g)$, the rate expression is:

rate = k (conc NO)^2 (conc O$_2$)

Show that the rate law is consistent with the mechanism

$NO(g) + O_2(g) \rightleftharpoons NO_3(g)$ fast

$NO_3(g) + NO(g) \rightarrow 2NO_2(g)$ slow

Exercise 16.3   Answer

The second reaction is the rate determining step and is equal to the rate of the overall reaction. The rate of the second reaction is

rate = k' (conc NO$_3$) (conc NO) = rate of overall reaction

The equilibrium expression for the fast reaction is

$$K_c = \frac{(\text{conc NO}_3)}{(\text{conc NO})(\text{conc O}_2)} \quad \text{or} \quad (\text{conc NO}_3) = K_c\,(\text{conc NO})(\text{conc O}_2)$$

The conc of the unstable, high energy intermediate, NO_3, can be eliminated from the rate expression for the second reaction by substituting for the conc NO_3.

rate = k' (conc NO)(K_c)(conc NO)(conc O_2)

Taking k'K_c to equal k, we have

rate = k (conc NO)^2(conc O_2)

The proposed mechanism agrees with the rate expression. The fact that a mechanism is consistent with the rate expression does not prove that the mechanism is correct. Notice also that the proposed mechanism involves successive collisions between two molecules even though the reaction is third order overall. Effective three body collisions seldom occur. A reaction that involves the rapid establishment of equlibrium, followed by a slower, rate determining step is a more reasonable mechanism for third order reactions.

SUGGESTED ASSIGNMENT XVI

Text: Chapter 16

Problems: Numbers 16.1–16.5, 16.22–16.24, 16.26–16.28, 16.32, 16.35

Solutions to Assigned Problems Set XVI

16.1

$$\text{rate} = \frac{-\Delta \text{conc } CH_3CHO}{\Delta t} = \frac{-(0.0248 - 0.0266)M}{5.4 \text{ min}} = \frac{-(-0.0018 \text{ M})}{5.4 \text{ min}}$$

$= 3.3 \times 10^{-4}$ mole liter^{-1} min^{-1}

16.2

a. The overall order of reaction is equal to the sum of the exponents. Since this reaction is first order with respect to both CO and NO_2, the overall order of reaction is $1 + 1 = 2$.

b. rate = k (conc CO)(conc NO_2)

$\quad$ = (0.50 liter mole^{-1} sec^{-1})(0.025 mole liter^{-1})(0.040 mole liter^{-1})

$\quad$ = 5.0×10^{-4} mole liter^{-1} sec^{-1}

16.3

Rearrange Equation 16.5 and solve for t.

$$t = \left(\frac{2.30}{k}\right) \log_{10} \frac{X_0}{X} = \frac{2.30}{0.35} \log_{10} \frac{0.160}{0.020} = 6.57 \log_{10} 8.0;$$

$\log_{10} 8.0 = 0.90$, so $t = (6.57)(0.90) = 5.9$ min

16.4

Apply Equation 16.10 with $T_2 = 800°K$, and $T_1 = 650°K$.

$$\log_{10} \frac{k_2}{k_1} = \frac{E_a}{(2.30)(1.99)} \left(\frac{T_2 - T_1}{T_2 T_1}\right)$$

$$\log_{10} \frac{23.0}{0.220} = \log_{10} 104 = \log_{10} (1.04 \times 10^2) = \log 1.04 + 2 = 0.017 + 2 = 2.017$$

$$2.017 = \frac{E_a (800 - 650)}{(2.30)(1.99)(800)(650)} = \frac{150 \, E_a}{2.38 \times 10^6};$$

$E_a = 3.20 \times 10^4$ cal $= 32.0$ kcal

16.5

We can write the rate for the slow reaction directly from the equation.

rate $= k_b$ (conc A_2)(conc B)

The concentration of B is governed by the position of the equlibrium in the first reaction.

$K_c = \frac{(\text{conc B})^2}{(\text{conc B}_2)}$, or (conc B) $= (K_c)^{1/2}$ (conc B_2)$^{1/2}$

Substituting this expression for the conc B in the rate expression:

rate $= k_b$ (conc A_2)$(K_c)^{1/2}$ (conc B_2)$^{1/2}$, which is equivalent to the given rate expression with $k = k_b (K_c)^{1/2}$

16.22

rate $= k$ (conc O_3) (conc $C_2 H_4$)
  $= (2 \times 10^3$ liter mole^{-1} sec^{-1})(5×10^{-8} mole liter^{-1}) $\times$ (1×10^{-8} mole liter^{-1})
  $= 1 \times 10^{-12}$ mole liter^{-1} sec^{-1}

time $= \frac{\text{conc}}{\text{rate}} = \frac{1 \times 10^{-8} \text{ mole/liter}}{1 \times 10^{-12} \text{ mole/liter sec}} = 1 \times 10^4$ sec (about 2.7 hrs)

16.23

a. The conc of B remains constant while the conc of A is increased by a factor of two from experiment two to experiment one. The rate increases by a factor of four. The reaction is second order with respect to A. Comparing the rate of experiment two to that of experiment three, the conc of A is held constant and the conc of B is doubled. This results in doubling the reaction rate. The reaction is first order with respect to B.

b. The rate expression is:

rate = k (conc A)^2 (conc B)

Using the data of experiment one:

$$k = \frac{\text{rate}}{(\text{conc A})^2 (\text{conc B})} = \frac{6.00 \times 10^{-3} \text{ mole liter}^{-1} \text{ sec}^{-1}}{(0.500 \text{ mole liter}^{-1})^2 (0.400 \text{ mole liter}^{-1})}$$

= 6.00 × 10^{-2} liter^2 mole^{-2} sec^{-1}

16.24

a. For zero order, rate = k = 0.0050 mole liter^{-1} sec^{-1}

b. For first order, rate = kX; k = $\frac{\text{rate}}{X}$ = $\frac{5.0 \times 10^{-3} \text{ mole liter}^{-1} \text{ sec}^{-1}}{2.00 \times 10^{-1} \text{ mole liter}^{-1}}$

= 2.5 × 10^{-2} sec^{-1}

c. For second order, rate = kX^2; k = $\frac{\text{rate}}{X^2}$ =

$\frac{5.0 \times 10^{-3} \text{ mole liter}^{-1} \text{ sec}^{-1}}{(2.00 \times 10^{-1} \text{ mole liter}^{-1})^2}$ = 1.2 × 10^{-1} liter mole^{-1} sec^{-1}

16.26

Prepare a table similar to that in Example 16.3.

t	X	$\log_{10} X$	$1/X$
0	0.600	−0.222	1.67
20	0.400	−0.398	2.50
40	0.267	−0.573	3.75
60	0.182	−0.740	5.49

The linear plot will be that of $\log_{10} X$ vs. t, a characteristic of a first order reaction.

$$\frac{\Delta \log \text{conc}}{\Delta t} \approx -0.0088 \approx -8.8 \times 10^{-3} \text{sec}^{-1} \approx \text{constant slope};$$

$\log_{10} \frac{X_0}{X} = \frac{kt}{2.30}$; but slope $= \frac{-k}{2.30}$ and $k = 2.30 \,(8.8 \times 10^{-3})\, \text{sec}^{-1}$

$= 2.0 \times 10^{-2} \text{sec}^{-1}$

16.27

a. For a first order reaction, $t_{1/2} = \frac{0.693}{k}$; (Equation 16.6)

$$k = \frac{0.693}{t_{1/2}} = \frac{0.693}{650 \text{ sec}} = 1.07 \times 10^{-3} \text{ sec}^{-1}$$

b. After 650 seconds the concentration is 0.0250 M; at the end of the next 650 seconds the concentration is 0.0125 M. The time required is $2 \times 650 \text{ sec} = 1300 \text{ sec}$.

c. Taking 0.0125 M as the initial concentration and 3600 sec as t, substitute into Equation 16.5, using the above value of k.

$$\log_{10} \frac{X_0}{X} = \frac{kt}{2.30} = \frac{(1.07 \times 10^{-3})(3600)}{2.30} = 1.67; \text{ the antilog of}$$

1.67 is 4.6×10^1; $\frac{0.0125}{X} = 4.6 \times 10^1$; $X = 2.71 \times 10^{-4}$ M

16.28

a. Consider the concentration of 0.0300 M at 200 minutes to be the initial concentration and the initial time, and substitute into Equation 16.5.

$$\log_{10} \frac{X_0}{X} = \frac{kt}{2.30}; \log_{10} \frac{0.0300}{0.0200} = \frac{(k)(200 \text{ min})}{2.30}; \log_{10} 1.5 = 0.176$$

$$k = \frac{(0.176)(2.30)}{200 \text{ min}} = 2.02 \times 10^{-3} \text{ min}^{-1}$$

b. Again, use Equation 16.5 and solve for X_0 using the calculated value of k.

$$\log_{10} \frac{X_0}{X} = \frac{kt}{2.30} \; ; \log_{10} \frac{X_0}{0.0200} = \frac{(2.02 \times 10^{-3})(400)}{2.30} = 0.351$$

The antilog of 0.351 is 2.24, so $\frac{X_0}{0.0200} = 2.24$; $X_0 = 0.0448$ M

16.32

First we need to calculate ΔH for the reaction. Recall that $\Delta H_f \, O_2$ is zero.

$\Delta H_r = 3\Delta H_f O_2(g) - 2\Delta H_f \, O_3(g) = 3(0) - 2(34 \text{ kcal}) = -68 \text{ kcal}$

$E'_a = E_a - \Delta H = 28 \text{ kcal} - (-68 \text{ kcal}) = 96 \text{ kcal}$

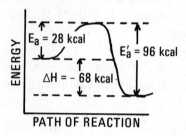

16.35

The sum of the two step mechanism gives the overall equation

$2O_3(g) \rightarrow 3O_2(g)$

We can write the rate expression in terms of the slow, or rate determining, step.

rate = k_2 (conc O_3) (conc O)

Atomic oxygen does not occur in the overall equation, but the concentration of O atoms is governed by the position of the equilibrium in the fast step.

$$K = \frac{(\text{conc } O_2)(\text{conc } O)}{(\text{conc } O_3)} \; ; \text{ or } (\text{conc } O) = \frac{K(\text{conc } O_3)}{(\text{conc } O_2)}$$

Substituting this expression for the conc O in the above rate expression gives:

$$\text{rate} = k_2 \, (\text{conc } O_3) \, \frac{K \, (\text{conc } O_3)}{(\text{conc } O_2)}$$

Let $k = k_2 K$, $\text{rate} = k \dfrac{(\text{conc } O_3)^2}{(\text{conc } O_2)}$

This rate law indicates that the decomposition of ozone is inversely proportional to the concentration of the product, O_2. The proposed mechanism and the derived rate law indicate that the reaction is not just a simple collision between two O_3 molecules as we might have supposed based on the overall equation for the reaction.

UNIT 17

The Atmosphere

The goal of this unit is to integrate the principles of chemical thermodynamics, equilibrium, kinetics, and chemical bonding as they apply to components of the atmosphere and the problems of air pollution.

INSTRUCTIONAL OBJECTIVES

17.1 The Composition of the Atmosphere
  *1. Convert gas phase concentrations given in mole per cent, volume per cent, weight per cent, partial pressure, moles per liter, ppm, or ppb to any of the other units. (Probs. 17.1, 17.22)
  2. List the major constituents of the atmosphere and describe how each can be extracted from the air.

17.2 Nitrogen
  1. Account for the lack of reactivity of elementary nitrogen; list some uses of nitrogen based on this inactivity.
  2. Describe the natural process of nitrogen fixation.
  3. Discuss the chemistry of the Haber process from the standpoint of equilibrium and rate considerations and from the economics of the cost of raw materials.
  4. Write balanced equations for the reaction of N_2 with active metals, H_2, and O_2.

17.3 Oxygen
  *1. Write equations for the formation of ionic oxides and for the reactions of basic anhydrides with water. (Prob. 17.2)
  2. Realize that in some transition metal oxides the bonding is predominately covalent.
  *3. Write balanced equations for the formation of peroxides and superoxides and for the reactions of these compounds with water. (Probs. 17.2, 17.24)

4. Recognize that bonding in nonmetal oxides is predominately covalent and that the higher oxide usually forms at low temperatures.
*5. Write balanced equations for the formation of nonmetal oxides and for the reactions of acid anhydrides with water. (Probs. 17.2, 17.24)

17.4 The Noble Gases
1. List some properties and uses of noble gases.
2. Realize that noble gas compounds have recently been prepared.

17.5, 17.6 Carbon Dioxide; Water Vapor
1. List some methods for the production of CO_2.
2. Discuss some possible effects of increasing concentrations of CO_2 in the atmosphere.
*3. Calculate the third quantity, given any two of the following: relative humidity, partial pressure of water in the air, equilibrium vapor pressure of water. (Prob. 17.3)
4. Discuss the principles of cloud seeding.

17.7 The Upper Atmosphere
1. Account for the presence of thermodynamically unstable species at high altitudes.
*2. Calculate the maximum wavelength of radiation that will initiate a given photochemical reaction. (Prob. 17.31)
*3. Write balanced equations to show the mechanisms of catalytic decomposition of ozone.
4. Discuss some possible causes and effects of a decrease in the concentration of the ozone layer.

17.8 Air Pollution
1. List five principal types of air pollutants, discuss their formation, their reactions in the atmosphere, and their effects on human beings and the environment.
2. Discuss several approaches to the reduction of hydrocarbons, carbon monoxide, and nitrogen oxides in automobile exhaust emissions.
3. Describe several methods for the reduction of oxides of sulfur and suspended particles in power plant and industrial emissions.

SUMMARY

1. Components of the Atmosphere

The composition of a gaseous mixture such as air can be expressed in various concentration units as indicated in Table 17.2 in the text. The major components in this mixture can be separated by the fractional distillation of liquid air. Elementary nitrogen, the most abundant component, is very unreactive due to the strength of the triple bond in the N_2 molecule. At high temperatures, nitrogen reacts with very reactive metals to form ionic nitrides, and forms covalently bonded nitrides with some of the less reactive metals.

Nitrogen fixation by bacteria is a vital natural process in the nitrogen cycle. The Haber process applies principles of chemical equilibrium, kinetics, and thermodynamics to the production of ammonia. The ammonia is used primarily for fertilizer, either directly or in the form of ammonium salts. Some ammonia is oxidized to produce nitric acid by the Ostwald process. Since hydrocarbons are the source of most of the hydrogen used in the Haber process, the cost of fertilizer has increased greatly within the last few years.

Oxygen reacts with most metals to form ionic oxides. These oxides (basic anhydrides) react with water to form basic solutions. A few transition metals form oxides that have a macromolecular structure in which the bonding is predominately covalent. Sodium and barium react with oxygen to form peroxides. Oxygen exhibits an oxidation number of -1 in these compounds. In superoxides such as KO_2, oxygen has an oxidation number of minus one-half. Both peroxides and superoxides react vigorously with water to produce hydrogen peroxide.

Nonmetals react with oxygen to produce compounds in which the bonding is predominately covalent. While SiO_2 and B_2O_3 are macromolecular, most nonmetal oxides are molecular. Many of these molecular species react with water to produce acidic solutions. Such nonmetallic oxides are known as acid anhydrides. Sulfur dioxide is one of the most important nonmetallic oxides since it is further oxidized to sulfur trioxide to produce sulfuric acid.

In reactions of oxygen with metals and with nonmetals, the formation of the higher oxide (i.e., the oxide containing the greatest percentage of oxygen) is favored at low temperatures and high concentrations of oxygen.

Stable compounds of noble gases have recently been prepared. However, uses of the noble gases are related to their almost complete lack of chemical reactivity. Although carbon dioxide is present in trace amounts in the atmosphere, its concentration has increased significantly in this century. This increase is primarily due to the combustion of fossil fuels. The "greenhouse effect" of CO_2 on the earth's temperature is another environmental concern.

The concentration of water vapor in the atmosphere varies considerably. The concentration of water vapor is often expressed in terms of relative humidity, which is related to the partial pressure of water vapor in the air and the equilibrium vapor pressure of water at the same temperature. This relationship is illustrated in Example 17.3 and Problem 17.3.

The upper atmosphere contains a variety of high energy, unstable species. These species form as a result of the interaction of high energy radiation with such molecular species as NO_2, O_2, and H_2O. (See Problems 17.13 and 17.31 in the text.) One of the most important species is ozone, an allotrope of oxygen. The concentration of ozone is relatively small and is found in a layer at an altitude of 20 km to 40 km. Ozone absorbs dangerous ultraviolet radiation from the sun. There has recently been much concern about the possible decrease in the concentration of ozone due to nitric oxide emissions from the supersonic transport aircraft. Halogenated hydrocarbons from aerosol propellants and refrigerants may also produce chlorine atoms, which serve as catalysts for the decomposition of ozone.

Exercise 17.1
1. A weather report gives the relative humidity at 75%, the temperature as 22.5°C, and the barometric pressure as 750 mmHg. (The equilibrium vapor pressure of water is 20 mmHg at 22.5°C.)
   a. What is the mole fraction of water vapor in the air?
   b. What is the concentration of water in the air in moles/liter?
   c. What is the weight percentage of water in the air?
   d. What is the concentration of water in ppm?
2. The compounds with formulas of SO_3, KO_2, and K_2O may be distinguished from each other by their behavior on addition to water. Describe in words and by means of chemical equations the reactions that occur when each substance is added to water.
3. A reaction in the formation of photochemical smog is: $NO_2(g) \rightarrow NO(g) + O(g)$; $\Delta H = 73$ kcal. Can this reaction be brought about by visible light?

Exercise 17.1 Answers
1. a. If the relative humidity is 75%, the partial pressure of the water vapor in the air is 0.75×20 mmHg = 15 mmHg.

$$X = \frac{\text{partial pressure}}{\text{total pressure}} = \frac{15 \text{ mmHg}}{750 \text{ mmHg}} = 0.020$$

   b. moles/liter = $\dfrac{\text{partial press in atm}}{(0.0821) T} = \dfrac{15/760}{(0.0821)(295)} = 8.1 \times 10^{-4}$ moles/liter

   c. The volume % = $100X = 100(0.020) = 2.0\%$

$$\text{Weight \%} = \text{volume \%} \times \frac{18}{19} = 2.0 \times \frac{18.0}{29.0} = 1.2\%$$

   d. ppm = $10^6 X = (10^6)(.020) = 2.0 \times 10^4$

2. $SO_3(g) + H_2O \rightarrow H_2SO_4(l)$

$2KO_2(s) + 2H_2O \rightarrow 2K^+(aq) + 2OH^-(aq) + H_2O_2(aq) + O_2(g)$

$K_2O(s) + H_2O \rightarrow 2K^+(aq) + 2OH^-(aq)$

The potassium compounds both form basic solutions. Gaseous oxygen is released by the superoxide. The SO_3 forms an acid solution.

3. The approach here is to calculate the maximum wavelength and compare this wavelength to wavelengths of visible light. Use Equation 17.33.

$$\lambda_{max} = \frac{2.86 \times 10^5}{\Delta E \text{ (kcal)}} = \frac{2.86 \times 10^5}{73} = 3.9 \times 10^3 \text{ Å}$$

The range of the visible spectrum is 4000 to 7000 Å. The maximum wavelength of 3900 Å is just at the edge of the visible spectrum.

2. Air Pollution

Of the five major air pollutants, oxides of sulfur and suspended particles are produced primarily by power generators and industrial processes. In addition to absorbing and scattering light, particulate matter from industrial processes can cause various types of lung disease. The oxides of sulfur are harmful due to their conversion to sulfuric acid. Suspended particles can be removed from stack gases by electrical precipitators. The problems associated with the removal of oxides of sulfur are somewhat more involved. Processes designed to remove SO_2 resulting from the combustion of fuels have not been very successful, nor have pre-combustion removal methods. Power plants have been restricted to the use of the more expensive and less plentiful low sulfur coal. This obviously is a short-term solution until a successful post-combustion process can be developed. Another approach is to convert high sulfur coal to cleaner burning lightweight hydrocarbons (coal gasification process).

Hydrocarbons, carbon monoxide, and oxides of nitrogen are major pollutants for which the principal source is the internal combustion engine. The oxides of nitrogen are involved in photochemical reactions that produce smog. These reactions produce ozone, a powerful oxidizing agent. The ozone in turn reacts with unburned hydrocarbons from engine exhausts to produce a variety of harmful products. In addition to hydrocarbons, carbon monoxide is produced by the automobile engine. The conversion of carbon monoxide to the dioxide in the atmosphere is a relatively slow reaction. A slight increase in the carbon monoxide concentration can have a significant effect on the oxygen-carrying capacity of the blood.

Although catalytic devices have been developed to help bring about a more complete conversion of hydrocarbons and carbon monoxide to water and carbon dioxide, problems associated with the production of oxides of nitrogen still remain.

SUGGESTED ASSIGNMENT XVII

Test: Chapter 17

Problems: Numbers 17.1, 17.2, 17.3, 17.22, 17.24, 17.25, 17.26, 17.28, 17.31, 17.34, 17.35

Solutions to Assigned Problems Set XVII

17.1

a. Volume % = 100X = 100 (9.34 × 10⁻³) = 9.34 × 10⁻¹

b. Weight % = volume % × $\dfrac{MW}{av.\ MW}$ = 9.34 × 10⁻¹ × $\dfrac{39.95}{29.0}$ = 1.29

c. Partial pressure = X (total pressure) = 9.34 × 10⁻³ (1 atm) = 9.34 × 10⁻³ atm

moles/liter = $\dfrac{\text{partial press (in atm)}}{0.0821\ (T)}$ = $\dfrac{9.34 \times 10^{-3}}{(0.0821)(298)}$ = 3.82 × 10⁻⁴

d. ppm = 10⁶ (X) = 10⁶ (9.34 × 10⁻³) = 9.34 × 10³

17.2

a. $4Al(s) + 3O_2(g) \rightarrow 2\ Al_2O_3(s)$

b. $Ba(s) + O_2(g) \rightarrow BaO_2(s)$

c. $BaO_2(s) + 2H_2O \rightarrow Ba^{2+}(aq) + 2OH^-(aq) + H_2O_2(aq)$

d. $CO_2(g) + H_2O \rightarrow H_2CO_3(aq)$

17.3

From Equation 17.32, the defining equation for relative humidity,
R. H. × $P^°_{H_2O}$ = P_{H_2O} = 0.52 × 17.5 mm Hg = 9.1 mmHg

17.22

Refer to Table 17.2 for the relationships among concentration units.

a. $X = \text{ppm}/10^6 = 11.2/10^6 = 1.12 \times 10^{-5}$

b. partial pressure $= X(\text{total pressure}) = 1.12 \times 10^{-5} \,(1 \text{ atm}) = 1.12 \times 10^{-5}$ atm

c. moles/liter $= \dfrac{\text{partial pressure in atm}}{0.0821\,(T)} = \dfrac{1.12 \times 10^{-5}}{(0.0821)(298)} = 4.58 \times 10^{-7}$

d. Using conversion factors,

$$4.58 \times 10^{-7}\,\dfrac{\text{mole}}{\text{liter}} \times \dfrac{28.0\text{g}}{\text{mole}} \times \dfrac{1 \text{ liter}}{1 \times 10^3 \text{cc}} \times \dfrac{1 \times 10^6 \text{cc}}{\text{m}^3} = 1.28 \times 10^{-2}\,\text{g/m}^3$$

17.24

a. $Na_2O_2(s) + 2H_2O \rightarrow 2Na^+(aq) + 2OH^-(aq) + H_2O_2(aq)$

b. This conversion involves the reactions in the Ostwald process. Refer to text Equations 17.6, 17.7, 17.8.

c. $2NH_3(g) + CO_2(g) \rightarrow CO(NH_2)_2(s) + H_2O(l)$

d. The oxide with the lower percentage of oxygen is formed at high temperature. Heat CuO to bring about the conversion.

$2CuO(s) \rightarrow Cu_2O(s) + \tfrac{1}{2}O_2(g)$

e. Refer to Equation 17.9

f. P_4O_{10} is the acid anhydride of H_3PO_4.

$P_4O_{10}(s) + 6H_2O \rightarrow 4H_3PO_4(aq)$

17.25

a. Increase the pressure to increase both the yield and the rate of reaction. Also, use a catalyst to increase the rate.

b. Same as in (a).

17.26

Review the rules for writing Lewis structures in Chapter 8.

a. $[:\ddot{\text{N}}:]^{3-}$

b. $[:\ddot{\text{O}}-\text{H}]^{-}$

c. 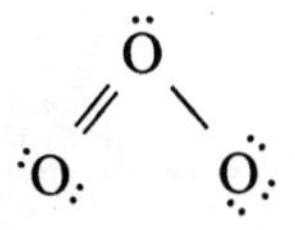
$$\begin{array}{c} \ddot{\text{O}} \\ \parallel \\ \cdot\ddot{\text{O}} \diagup \diagdown \ddot{\text{O}}: \end{array}$$

d. $:\ddot{\text{O}}\cdot$

17.28

The equilibrium reaction is

$$SO_2(g) + \tfrac{1}{2}O_2(g) \rightleftharpoons SO_3(g)$$

a. 100 − 20 = mole % O_2 + mole % SO_2 = 80; but mole % O_2 = ½ mole % SO_2. Let x = mole % SO_2; 0.50x = mole % O_2. 80 = x + 0.5x; x = 53% SO_2; 0.50 (53%) = 27% O_2; mole % SO_3 = 20; mole % SO_2 = 53%: mole % O_2 = 27%

b. $n = \dfrac{PV}{RT} = \dfrac{(1\ \text{atm})(1.0\ \text{liter})}{\left(0.0821\ \dfrac{\text{liter atm}}{°\text{K mole}}\right)(1073°\text{K})} = 1.14 \times 10^{-2}$ mole

c. moles/liter SO_3 = $(1.14 \times 10^{-2})(0.20) = 2.3 \times 10^{-3}$

   moles/liter SO_2 = $(1.14 \times 10^{-2})(0.53) = 6.0 \times 10^{-3}$

   moles/liter O_2 = $(1.14 \times 10^{-2})(0.27) = 3.1 \times 10^{-3}$

d. $K_c = \dfrac{[SO_3]}{[SO_2][O_2]^{\frac{1}{2}}} = \dfrac{2.3 \times 10^{-3}}{(6.0 \times 10^{-3})(3.1 \times 10^{-3})^{\frac{1}{2}}} = 6.9$

17.31

Use Equation 17.33 in the form

$$\Delta E = \dfrac{2.86 \times 10^5}{\lambda_{\text{max}}} = \dfrac{2.86 \times 10^5}{6800} = 42\ \text{kcal/mole}$$

Note that the visible spectrum ranges from about 4000 to 7000 Å. The wavelength here is much less energetic than the ultraviolet wavelengths listed in Table 17.7.

17.34

Substitute into the rate expression.

rate = k(conc CO)(conc NO)^2

$$= 3 \times 10^{-27} \frac{\text{liter}^2}{\text{mole}^2 \text{ sec}} \left(4 \times 10^{-6} \frac{\text{mole}}{\text{liter}}\right) \left(4 \times 10^{-9} \frac{\text{mole}}{\text{liter}}\right)^2$$

$$= 2 \times 10^{-49} \frac{\text{mole}}{\text{liter sec}}$$

This conversion of CO to CO_2 by NO would help solve two serious pollution problems caused by automobile engine emissions if the rate of this reaction were not so extremely slow.

17.35

The following conversion factors are needed.

1 km = 10^3 m; thus 1 km^3 = 10^9 m^3; 10^3 liter = 1 m^3

a. 6000 km^3 × $\frac{10^9 \text{ m}^3}{\text{km}^3}$ × $\frac{10^3 \text{ liter}}{1 \text{ m}^3}$ × 1 × 10^{-8} $\frac{\text{mole}}{\text{liter}}$ × $\frac{100 \text{g}}{\text{mole}}$

= 6 × 10^9 grams

b. The 1976 emission standard for hydrocarbons is 0.4 gram/mile.

6 × 10^9 grams × $\frac{1 \text{ mile}}{0.4 \text{ gram}}$ = 1.5 × 10^{10} miles

UNIT 18

Precipitation Reactions

The goal of this unit is to develop principles to predict the conditions under which precipitation reactions will occur, predict the products of such reactions, and to write chemical equations to represent these reactions.

INSTRUCTIONAL OBJECTIVES

18.1   Net Ionic Equations
 *1.   Write net ionic equations to represent precipitation reactions. (Probs. 18.1, 18.26)
 *2.   Based on an equation for a precipitation reaction and given the quantities of reactants, perform calculations to determine:
   a. The number of moles (or grams) of solid formed.
   b. The number of moles (or grams) of each ion remaining in solution. (Prob. 18.2)
   c. The concentration of each ion remaining in solution.

18.2   Solubilities of Ionic Compounds
   1. Be familiar with the solubility rules — Table 18.1 in the text.
 *2.   Predict if precipitation will occur when 0.1 M solutions of two different ionic compounds are mixed. (Probs. 18.1, 18.26)
   3. Discuss the factors that determine the solubility of an ionic compound.

18.3   Solubility Equilibria
   1. Describe the common ion effect.
   2. State the solubility product principle.

*3. Write the expression for K_{sp} when given the formula of an ionic compound.
*4. Given either of the following, calculate the other:
   a. The solubility of an ionic compound. (Prob. 18.3a)
   b. K_{sp}. (Prob. 18.3b)
*5. Calculate the equilibrium concentration of an ion in solution (e.g., OH^-), given the concentration of the other ion (e.g., Mg^{2+}) and the K_{sp} value. (Prob. 18.4a)
*6. Determine whether or not a precipitate will form upon mixing two solutions, given the K_{sp} value and the volumes and concentrations of the solutions before mixing. (Probs. 18.4b, 18.34)
*7. Calculate the percentage of an ion remaining in solution, given the K_{sp} value and the conditions of precipitation.
*8. Decide which of two possible precipitates will form when two solutions are mixed, given the K_{sp} values and the volumes and concentrations of the solutions before mixing. (Prob. 18.37)

18.4, 18.5 Applications of Precipitation Reactions
*1. Calculate the percentage of a component present in a mixture, given analytical data for a precipitation reaction. (Probs. 18.5, 18.40)
*2. Devise a scheme to separate and identify ions in a mixture by use of the solubility rules. (Prob. 18.41)
*3. Outline methods of preparing a variety of soluble and insoluble electrolytes by use of precipitation reactions. (Prob. 18.6)

SUMMARY

1. Solubilities of Ionic Compounds

A precipitation reaction will occur when solutions of electrolytes are mixed if one of the possible products is insoluble. The solubility rules (Table 18.1) can be used to predict the resulting precipitation reactions. For example, suppose a test tube containing a solution of $AgNO_3$ is added to a beaker that contains a solution of Na_2CO_3. According to the solubility rules, a precipitate of Ag_2CO_3 should be formed. The ionic species that were present in the test tube were $Ag^+(aq)$ and $NO_3^-(aq)$. The ionic species in the beaker before mixing were $Na^+(aq)$ and $CO_3^{2-}(aq)$. When the contents of the test tube were added to the beaker, a reaction occured to produce Ag_2CO_3. This reaction can be represented by the following balanced equation.

$$2Ag^+(aq) + CO_3^{2-}(aq) \rightarrow Ag_2CO_3(s)$$

Note that the $Na^+(aq)$ and $NO_3^-(aq)$ play no apparent part in this reaction. This is a net ionic equation. Net ionic equations contain only the predominant reacting species.

Exercise 18.1
1. Write net ionic equations for the reactions, if any, that occur when 0.1 M solutions of the following ionic compounds are mixed.
   a. K_2SO_4 and $Pb(NO_3)_2$
   b. NH_4CO_3 and $CaCl_2$
   c. $Al_2(SO_4)_3$ and KOH
   d. $Fe_2(SO_4)_3$ and $Sr(OH)_2$
2. When 75 ml of 0.20 M Na_2S solution is added to 45 ml of 0.40 M $AgNO_3$ solution, a black precipitate forms.
   a. Write a net ionic equation for the reaction.
   b. Calculate the number of grams of precipitate formed.
   c. Calculate the molarity of the Na^+ and the S^{2-} remaining in solution after precipitation.

Exercise 18.1 Answers
1. Refer to the solubility rules to decide if an insoluble compound will form.
   a. $PbSO_4$ precipitates, KNO_3 is soluble.

   $$Pb^{2+}(aq) + SO_4^{2-}(aq) \rightarrow PbSO_4(s)$$

   b. Possible precipitates: NH_4Cl and $CaCO_3$. Only the $CaCO_3$ precipitates.

   $$Ca^{2+}(aq) + CO_3^{2-}(aq) \rightarrow CaCO_3(s)$$

   c. Possible precipitates: K_2SO_4 and $Al(OH)_3$. Only $Al(OH)_3$ precipitates.

   $$Al^{3+}(aq) + 3OH^-(aq) \rightarrow Al(OH)_3(s)$$

   d. Possible precipitates: $SrSO_4$ and $Fe(OH)_3$. Both are insoluble. Two reactions occur that produce precipitates.

   $$Fe^{3+}(aq) + 3OH^-(aq) \rightarrow Fe(OH)_3(s)$$

   $$Sr^{2+}(aq) + SO_4^{2-}(aq) \rightarrow SrSO_4(s)$$

2. a. There are two possible precipitates: $NaNO_3$ and Ag_2S. All nitrates are soluble, so the equation is

   $$2Ag^+(aq) + S^{2-}(aq) \rightarrow Ag_2S(s)$$

b. Start by calculating the number of moles of Ag^+ and S^{2-} available for reaction.

$$0.075 \, \ell \times \frac{0.20 \text{ moles } S^{2-}}{\ell} = 0.015 \text{ mole } S^{2-}$$

$$0.045 \, \ell \times \frac{0.40 \text{ moles } Ag^+}{\ell} = 0.018 \text{ moles } Ag^+$$

But which ion is in excess for the reaction? Note that for every mole of S^{2-} present, 2 moles of Ag^+ react. The limiting factor is the number of moles of Ag^+ available. The number of grams of Ag_2S formed is

$$0.018 \text{ mole } Ag^+ \times \frac{1 \text{ mole } Ag_2S}{2 \text{ mole } Ag^+} \times \frac{248g}{\text{mole } Ag_2S} = 2.2g \; Ag_2S$$

c. To calculate the molarity of the Na^+, first determine the number of moles present. Do not forget that this number of moles is now present in 75ml + 45ml = 120ml or 0.120 liters.

$$0.075 \, \ell \times \frac{0.20 \text{ mole } Na_2S}{\ell} \times \frac{2 \text{ mole } Na^+}{1 \text{ mole } Na_2S} \times \frac{1}{0.120 \, \ell}$$

$$= 0.25 \; \frac{\text{mole } Na^+}{\ell}$$

To calculate the concentration of the S^{2-} remaining, determine the number of moles that are in excess.

$$0.018 \text{ moles } Ag^+ \times \frac{1 \text{ mole } S^{2-}}{2 \text{ mole } Ag^+} = 0.0090 \text{ moles } S^{2-} \text{ reacted.}$$

0.015 mole − 0.0090 moles = 0.006 mole S^{2-} remaining unreacted.

The molarity is $\dfrac{0.006 \text{ mole } S^{2-}}{0.120 \, \ell} = 0.05 \, \dfrac{\text{mole } S^{2-}}{\ell}$

2. Solubility Equilibria

When the silver carbonate discussed earlier is precipitated by the reaction of silver ions with carbonate ions, the ionic crystals of Ag_2CO_3 are in equilibrium with Ag^+ and CO_3^{2-} in solution. The same equilibrium is established when solid Ag_2CO_3 is added to water and stirred. A saturated solution is obtained. This equilibrium may be represented by the equation

$$Ag_2CO_3(s) \rightleftharpoons 2Ag^+(aq) + CO_3^{2-}(aq)$$

In order to deal quantitatively with precipitation reactions, it is necessary to apply the concept of the equilibrium constant to the equilibrium of saturated solutions. Using the Ag_2CO_3 equilibrium, $Ag_2CO_3(s) \rightleftharpoons 2Ag^+(aq) + CO_3^{2-}(aq)$, the equilibrium constant expression is $K_{sp} = [Ag^+]^2 [CO_3^{2-}] = 1.3 \times 10^{-11}$. Note that solid Ag_2CO_3 does not appear in the expression. The brackets indicate equilibrium concentrations in moles/liter.

When silver carbonate dissolves in pure water to form a saturated solution, the concentration of Ag^+ is twice the concentration of the CO_3^{2-}. The relative concentrations of these two ions in equilibrium with solid silver carbonate may be altered. The addition of silver ions from a more soluble silver salt such as silver nitrate will increase the concentration of silver ions. The added silver ions will combine with the carbonate ions in solution to precipitate additional silver carbonate and reduce the equilibrium concentration of carbonate ions. This behavior is referred to as the common ion effect. The addition of sodium sulfide to the silver carbonate equilibrium system will also alter the relative concentrations of the silver and carbonate ions. In this case, very insoluble silver sulfide forms, resulting in a decrease in the concentration of the silver ion. More silver carbonate dissolves and the concentration of the carbonate ion increases.

The solubility product principle that has been discussed is strictly applicable only in very dilute solutions. Interionic attractions cause deviations from ideal behavior.

Problems involving K_{sp} can be classified as follows:

1. *Calculation of a solubility from the K_{sp} value.* This is illustrated in Example 18.3b. Note that the relationship between K_{sp} and solubility is dependent upon the type of salt.

2. *Calculation of K_{sp} from solubilities.* This is shown in Example 18.3a. Be sure to employ the correct mole relationships in determining the concentrations of ions.

3. *Predicting whether or not a precipitate will form when two solutions are mixed.* This type of problem is shown in Example 18.5. You must consider the effect of dilution when a final solution is formed by mixing two solutions. Determine the concentrations of ions in the final solution, then compare the ion product to the K_{sp} value. Remember: ion product > K_{sp}, precipitate forms; ion product < K_{sp}, no precipitate forms.

4. *Determine the equilibrium concentration of an ion in solution,*

when given the concentration of the other ion. This type of problem is straightforward, as shown in the following example. Calculate the concentration of Ag^+ in equilibrium with 0.095M CrO_4^{2-}. The K_{sp} for Ag_2CrO_4 is 1.9×10^{-12}.

Substitute into the K_{sp} expression:

$$[Ag^+]^2 [CrO_4^{2-}] = 1.9 \times 10^{-12} = [Ag^+]^2 [0.095]$$

$$[Ag^+]^2 = \frac{1.9 \times 10^{-12}}{9.5 \times 10^{-2}} = 2.0 \times 10^{-11} = 20 \times 10^{-12}$$

$$[Ag^+] = 4.5 \times 10^{-6}$$

5. *Determine which of two possible precipitates will form when two solutions are mixed.* This type of problem has application for the separation of ions by selective precipitation. It is the basis for some of the separations in qualitative analysis. To illustrate the principles that apply in this type of problem, consider the following example: Solid lead nitrate crystals are slowly added to a solution that is 0.10 M in I^- and 0.10 M in SO_4^{2-}. Which precipitate forms first, PbI_2 or $PbSO_4$? The K_{sp} value for PbI_2 is 1.4×10^{-8}; for $PbSO_4$ it is 1.6×10^{-8}. Begin by calculating the concentrations of Pb^{2+} required to start to precipitate PbI_2 and $PbSO_4$.

$$\text{For } PbI_2: [Pb^{2+}] = \frac{K_{sp}}{[I^-]^2} = \frac{1.4 \times 10^{-8}}{(0.10)^2} = 1.4 \times 10^{-6}$$

$$\text{For } PbSO_4: [Pb^{2+}] = \frac{K_{sp}}{[SO_4^{2-}]} = \frac{1.6 \times 10^{-8}}{0.10} = 1.6 \times 10^{-7}$$

The $PbSO_4$ will precipitate first; a smaller Pb^{2+} concentration is needed to precipitate $PbSO_4$ than to precipiate PbI_2. This problem may be extended as follows: Calculate the concentration of the first ion precipitated when the second precipitate starts to form.

The PbI_2 starts to precipitate when the $[Pb^{2+}] = 1.4 \times 10^{-6}$. The concentration of the SO_4^{2-} when $[Pb^{2+}] = 1.4 \times 10^{-6}$ is

$$[SO_4^{2-}] = \frac{K_{sp}}{[Pb^{2+}]} = \frac{1.6 \times 10^{-8}}{1.4 \times 10^{-6}} = 1.1 \times 10^{-2}$$

Note that this indicates that the concentration of SO_4^{2-} has been decreased from 0.10 M to 0.01 M before any PbI_2 begins to precipitate. Is the precipitation by the use of Pb^{2+} a practical method of separating SO_4^{2-} from I^-?

Exercise 18.2

1. Write the K_{sp} expression for
    a. AgBr   b. Ag_2S c. $Fe(OH)_3$ d. $Mg_3(PO_4)_2$

2. The K_{sp} for Ag_2CO_3 is 1.30×10^{-11}. Calculate the solubility of Ag_2CO_3 in moles/liter.
3. Calculate the equilibrium concentration of Ag^+ when the $[CO_3^{2-}] = 3.25 \times 10^{-4}$.
4. Will a precipitate of Ag_2CO_3 form when 125 ml of 2.00×10^{-2} M $AgNO_3$ solution is added to 175 ml of 1.40×10^{-1} M K_2CO_3 solution?
5. The solubility of AgCl is found to be 1.81×10^{-3} g/l. Calculate the K_{sp} value for AgCl.

Exercise 18.2 Answers
1. a. $K_{sp} = [Ag^+][Br^-]$
   b. $K_{sp} = [Ag^+]^2[S^{2-}]$
   c. $K_{sp} = [Fe^{3+}][OH^-]^3$
   d. $K_{sp} = [Mg^{2+}]^3[PO_4^{3-}]^2$
2. $K_{sp} = 1.30 \times 10^{-11} = [Ag^+]^2[CO_3^{2-}]$
   From the equation $Ag_2CO_3(s) \rightarrow 2Ag^+(aq) + CO_3^{2-}(aq)$ it is seen that for every mole of Ag_2CO_3 that dissolves, two moles of Ag^+ and one mole of CO_3^{2-} enter the solution. If S = solubility of Ag_2CO_3 in units of moles/liter, then $[CO_3^{2-}] = S$, and $[Ag^+] = 2S$

   Substituting into the K_{sp} expression:

   $(2S)^2(S) = 1.30 \times 10^{-11}$; $4S^3 = 1.30 \times 10^{-11}$; $S^3 = 3.26 \times 10^{-12}$

   Solving: $S = (3.25)^{1/3} \times 10^{-4} = 1.48 \times 10^{-4}$

3. $[Ag^+]^2[CO_3^{2-}] = 1.30 \times 10^{-11}$; $[Ag^+]^2 = \dfrac{1.30 \times 10^{-11}}{3.25 \times 10^{-4}} = 4.00 \times 10^{-8}$; $[Ag^+] = 2.00 \times 10^{-4}$

4. Calculate the concentrations of Ag^+ and CO_3^{2-} after the two solutions are mixed. Then compare the ion concentration product to the K_{sp} value. The volume of the solution after mixing is:

   125ml + 175 ml = 300 ml = 0.300 liter

   conc Ag^+: $0.125 \ell \times 2.00 \times 10^{-2} \dfrac{\text{mole}}{\ell} \times \dfrac{1}{0.300 \ell} = 8.33 \times 10^{-3}$

   conc CO_3^{2-}: $0.175 \ell \times 1.40 \times 10^{-1} \dfrac{\text{mole}}{\ell} \times \dfrac{1}{0.300 \ell} = 8.17 \times 10^{-2}$

   ion concentration product: $(8.33 \times 10^{-3})^2 (8.17 \times 10^{-2})$
   $= 5.67 \times 10^{-6}$

The ion concentration product is much greater than the K_{sp} value. Ag_2CO_3 will precipitate.

5. The solubility in moles per liter is:

$$1.81 \times 10^{-3} \frac{g}{\ell} \times \frac{1 \text{ mole}}{143g} = 1.27 \times 10^{-5} \text{ mole}/\ell$$

From the formula AgCl, it is evident that for every mole of AgCl that dissolves, a mole of Ag^+ and a mole of Cl^- enter the solution. The equilibrium concentrations of Ag^+ and Cl^- must each be 1.27×10^{-5} M.

$$K_{sp} = [Ag^+][Cl^-] = (1.27 \times 10^{-5}) \times (1.27 \times 10^{-5})$$
$$= 1.6 \times 10^{-10}$$

3. Application of Precipitation Reactions

Many procedures in analytical chemistry and a variety of synthesis processes in the chemical industry are based on principles of solubility equilibria. The following exercise illustrates these applications.

Exercise 18.3
1. Exactly 0.248g of dried Ag_2CrO_4 is obtained from a 1.648g sample of an alloy. Calculate the percentage of Ag in the alloy.
2. It requires 19.8 ml of 0.186 M $AgNO_3$ solution to precipitate all of the I^- in a 0.842g sample containing iodide ions. Determine the percentage of I^- in the sample.
3. What reagent should be used to convert:
    a. $CoSO_4$ to $CoCl_2$
    b. Na_2S to NaCl
    c. NaOH to $NaNO_3$

Exercise 18.3 Answers
1. This is an example of a gravimetric method of anlaysis. Relate grams of Ag_2CrO_4 to grams of silver using conversion factors.

$$0.248g \ Ag_2CrO_4 \times \frac{1 \text{ mole } Ag_2CrO_4}{332g} \times \frac{2 \text{ mole Ag}}{1 \text{ mole } Ag_2CrO_4} \times$$

$$\frac{108g}{\text{mole Ag}} \times \frac{1}{1.648g} \times 100 = 9.79\%$$

2. This is an example of a volumetric method of analysis. The equation for the reaction is: $Ag^+(aq) + I^-(aq) \rightarrow AgI(s)$

$$0.0198 \ \ell \times \frac{0.186 \text{ mole } AgNO_3}{\ell} \times \frac{1 \text{ mole } I^-}{1 \text{ mole } AgNO_3} \times \frac{127g}{\text{mole } I^-} \times$$

$$\frac{1}{0.842g} \times 100 = 55.5\%$$

3. a. Add $BaCl_2$, filter to remove solid $BaSO_4$, and evaporate the filtrate to give $CoCl_2$ crystals.
   b. Add $CuCl_2$, filter to remove solid CuS, evaporate the filtrate.
   c. Add $Cu(NO_3)_2$, filter to remove $Cu(OH)_2$, evaporate the filtrate.

SUGGESTED ASSIGNMENT XVIII

Text: Chapter 18

Problems: Numbers 18.1–18.6, 18.26, 18.31, 18.34, 18.37, 18.40, 18.41

Solutions to Assigned Problems Set XVIII

18.1

Possible precipitates are $BaCl_2$ and $Fe(OH)_2$. According to Table 18.1, $BaCl_2$ is soluble. Since Fe is not a 1A element, $Fe(OH)_2$ is insoluble.

Fe^{2+} (aq) + $2OH^-$(aq) → $Fe(OH)_2$ (s)

18.2

Use conversion factors to calculate the number of moles of $Ba(OH)_2$ and $FeCl_2$ present before reaction.

$$0.200 \text{ liter} \times \frac{0.10 \text{ mole Ba(OH)}_2}{\text{liter}} = 0.020 \text{ mole Ba(OH)}_2$$

$$0.300 \text{ liter} \times \frac{0.20 \text{ mole FeCl}_2}{\text{liter}} = 0.060 \text{ mole FeCl}_2$$

Based on the formulas, use conversion factors to calculate the number of moles of each ion present before reaction.

$$0.020 \text{ mole Ba(OH)}_2 \times \frac{1 \text{ mole Ba}^{2+}}{\text{mole Ba(OH)}_2} = 0.020 \text{ mole Ba}^{2+}$$

$$0.020 \text{ mole Ba(OH)}_2 \times \frac{2 \text{ mole OH}^-}{\text{mole Ba(OH)}_2} = 0.040 \text{ mole OH}^-$$

$$0.060 \text{ mole FeCl}_2 \times \frac{1 \text{ mole Fe}^{2+}}{\text{mole FeCl}_2} = 0.060 \text{ mole Fe}^{2+}$$

$$0.060 \text{ mole FeCl}_2 \times \frac{2 \text{ mole Cl}^-}{\text{mole FeCl}_2} = 0.12 \text{ mole Cl}^-$$

Based on the mole relationships in the equation for Problem 18.1, the

OH^- is the limiting reagent. The number of moles of Fe^{2+} that react with 0.040 mole OH^{-1} is

$$0.040 \text{ mole } OH^- \times \frac{1 \text{ mole } Fe^{2+}}{2 \text{ mole } OH^-} = 0.020 \text{ mole } Fe^{2+}$$

The number of moles of Fe^{2+} remaining = (0.060 - 0.020) mole = 0.040 mole Fe^{2+}. All of the OH^- has reacted and none of the Ba^{2+} and Cl^- has done so.

	Number of moles		
	original	change	final
Ba^{2+}	0.020	0	0.020
OH^-	0.040	-0.040	0.000
Fe^{2+}	0.060	-0.020	0.040
Cl^-	0.12	0	0.12

18.3

a. The equation for dissolving $CaCO_3$ is

$$CaCO_3(s) \rightarrow Ca^{2+}(aq) + CO_3^{2-}(aq)$$

Let S = solubility of $CaCO_3$; therefore, $[Ca^{2+}] = [CO_3^{2-}] = S$

$K_{sp} = [Ca^{2+}][CO_3^{2-}] = 5 \times 10^{-9} = (S)(S) = S^2$;

$S^2 = 50 \times 10^{-10}$; $S = 7 \times 10^{-5}$

The solubility of $CaCO_3$ in water is 7×10^{-5} mole/liter.

b. The equation is $Zn(OH)_2(s) \rightleftarrows Zn^{2+}(aq) + 2OH^-(aq)$. The $[Zn^{2+}] = 2.3 \times 10^{-6}$; $[OH^-] = 2(2.3 \times 10^{-6}) = 4.6 \times 10^{-6}$

Note that two moles of OH^- are produced for every mole of $Zn(OH)_2$ that dissolves.

$K_{sp} = [Zn^{2+}][OH^-]^2 = (2.3 \times 10^{-6})(4.6 \times 10^{-6})^2 = 4.9 \times 10^{-17}$

18.4

a. The equation is $CdS(s) \rightleftarrows Cd^{2+}(aq) + S^{2-}(aq)$

$$K_{sp} = [Cd^{2+}][S^{2-}]; [S^{2-}] = \frac{K_{sp}}{[Cd^{2+}]} = \frac{1 \times 10^{-27}}{1.0 \times 10^{-1}} = 1 \times 10^{-26}$$

b. We need to compare the ion concentration product to the K_{sp} in order to determine if a precipitate will form.

$$(\text{conc } Cd^{2+})(\text{conc } S^{2-}) = (1.0 \times 10^{-2})(1 \times 10^{-21}) = 1 \times 10^{-23}$$

Since $1 \times 10^{-23} > 1 \times 10^{-27}$, precipitation will occur until the concentrations of Cd^{2+} and S^{2-} decrease to the point at which the product of the concentrations of the ions becomes equal to 1×10^{-27}.

18.5

Again, we use conversion factors to calculate the number of grams of Ba in the sample.

$$0.466\text{g BaSO}_4 \times \frac{137\text{g Ba}}{233\text{g BaSO}_4} = 0.274\text{g Ba}$$

$$\frac{0.274\text{g}}{1.000\text{g}} \times 100 = 27.4\% \text{ Ba}$$

18.6

To convert $MgCl_2$ to $Mg(NO_3)_2$, add $AgNO_3$ to precipitate AgCl, filter off the AgCl, and evaporate the filtrate to obtain $Mg(NO_3)_2$.

To obtain KCl from K_2SO_4, add an equivalent amount of $BaCl_2$ to precipitate $BaSO_4$. Filter off the $BaSO_4$ and evaporate the filtrate to obtain the KCl.

18.26

Use Table 18.1 if you have not memorized the solubility rules.

a. Since sodium compounds are soluble, and Ba^{2+} is not a 1A element, the precipitate is $BaCO_3$.

$$Ba^{2+}(aq) + CO_3^{2-}(aq) \rightarrow BaCO_3(s)$$

b. Two precipitation reactions occur.

$$Ba^{2+}(aq) + SO_4^{2-}(aq) \rightarrow BaSO_4(s)$$

$Zn^{2+}(aq) + S^{2-}(aq) \rightarrow ZnS(s)$

c. All nitrates are soluble as are all Group 1A chlorides; therefore, there is no reaction.

d. $2Sb^{3+}(aq) + 3S^{2-}(aq) \rightarrow Sb_2S_3(s)$

18.31

Since $[Pb^{2+}] \times [Cl^-]^2 = K_{sp}\ PbCl_2 = 1.7 \times 10^{-5}$

$[Pb^{2+}] = \dfrac{1.7 \times 10^{-5}}{(2.0)^2} = 4.2 \times 10^{-6}$ mole/liter

18.34

a. We need to compare the ion concentration product to the K_{sp}.

$(\text{conc } Pb^{2+})(\text{conc } I^-)^2 = (1.0 \times 10^{-3})(2.0 \times 10^{-2})^2 = 4.0 \times 10^{-7}$

The K_{sp} for PbI_2 is 1×10^{-8}; precipitation will occur.

b. The reaction is $Ba^{2+}(aq) + 2F^-(aq) \rightarrow BaF_2(s)$

$K_{sp} = [Ba^{2+}][F^-]^2 = 2 \times 10^{-6}$

This problem is similar to 18.34(a) except that we need to calculate the concentrations of Ba^{2+} and F^- after mixing, but before any reaction. The volume after mixing is (500 + 300) ml = 800 ml

conc Ba^{2+} = 0.500 liter $\times$ 0.010 $\dfrac{\text{mole } Ba^{2+}}{\text{liter}} \times \dfrac{1}{0.800 \text{ liter}} = 6.25 \times 10^{-3}$ M

conc F^- = 0.300 liter $\times$ 0.010 $\dfrac{\text{mole } F^-}{\text{liter}} \times \dfrac{1}{0.800 \text{ liter}} = 3.75 \times 10^{-3}$ M

$(\text{conc } Ba^{2+})(\text{conc } F^-)^2 = (6.25 \times 10^{-3})(3.75 \times 10^{-3})^2 = 8.8 \times 10^{-8}$

Since $8.8 \times 10^{-8} < 2 \times 10^{-6}$, no precipitate will form.

18.37

Calculate the $[Ag^+]$ required to precipitate AgCl. For AgCl, $K_{sp} = 1.6 \times 10^{-10} = [Ag^+][Cl^-]$

$$[Ag^+] = \frac{K_{sp}}{[Cl^-]} = \frac{1.6 \times 10^{-10}}{1.0 \times 10^{-1}} = 1.6 \times 10^{-9}$$

The $[Ag^+]$ required to precipitate Ag_2CrO_4 is calculated in a similar way.

$$K_{sp} = 2 \times 10^{-12} = [Ag^+]^2[CrO_4^{2-}]$$

$$[Ag^+]^2 = \frac{K_{sp}}{[CrO_4^{2-}]} = \frac{2 \times 10^{-12}}{1.0 \times 10^{-1}} = 2 \times 10^{-11}$$

$$[Ag^+] = (20 \times 10^{-12})^{1/2} = 4.5 \times 10^{-6}$$

The AgCl precipitates first. The Ag_2CrO_4 starts to precipitate when the $[Ag^+] = 4.5 \times 10^{-6}$. The $[Cl^-]$ when the Ag_2CrO_4 starts to precipitate is

$$[Cl^-] = \frac{K_{sp}}{[Ag^+]} = \frac{1.6 \times 10^{-10}}{4.5 \times 10^{-6}} = 3.6 \times 10^{-5}$$

The chromate starts to precipitate when virtually all of the Cl^- has precipitated. Since Ag_2CrO_4 is bright red in color, while AgCl is white, CrO_4^{2-} is a suitable indicator for the titration of Cl^- with Ag^+.

18.40

The equation for the reaction is

$$Mg^{2+}(aq) + 2OH^-(aq) \rightarrow Mg(OH)_2(s)$$

The number of moles of Mg^{2+} present is

$$0.0214 \text{ liter} \times \frac{0.100 \text{ moles } OH^-}{\text{liter}} \times \frac{1 \text{ mole } Mg^{2+}}{2 \text{ mole } OH^-}$$

$$= 1.07 \times 10^{-3} \text{ mole } Mg^{2+}$$

This is the number of moles of Mg^{2+} in a 20.8 ml sample.

The concentration of $Mg^{2+} = \dfrac{1.07 \times 10^{-3} \text{ mole}}{0.0208 \text{ liter}} = 5.14 \times 10^{-2}$ M

18.41

Using the solubility rules, we want to select a reagent that will precipitate one ion and leave the other ions in solution. After removing the precipitate, we select another reagent to form a precipitate with one of the remaining ions.

a. Add Cl^- to precipitate Pb^{2+} as $PbCl_2$, leaving Ca^{2+} and Cu^{2+}. The Cu^{2+} can be removed by treating with S^{2-} to form CuS. The presence of Ca^{2+} can be shown by precipitation with CO_3^{2-} or SO_4^{2-} after the Pb^{2+} and Cu^{2+} have been removed.

b. Add Mg^{2+} to precipitate OH^- as $Mg(OH)_2$. Add Cu^{2+} to remove the S^{2-}. With the OH^- and S^{2-} separated, add Ag^+ to show the presence of Cl^-.

UNIT 19

Acids and Bases

The goal of this unit is to become familiar with properties of acidic and basic solutions, and the methods by which such solutions are formed. The relative strengths of acids and bases will be considered quantitatively and then explained on the basis of molecular structure.

INSTRUCTIONAL OBJECTIVES

19.1 The Dissociation of Water; Nature of Acids and Bases
 *1. Write the chemical equation for the dissociation of water, the expression for K_w,

 $$[H^+] \times [OH^-] = 1.0 \times 10^{-14}$$

 2. Relate acidic and basic properties of solutions to the relative concentrations of H^+ and OH^-. (Probs. 19.29, 19.30)
 *3. Calculate $[H^+]$ when given the $[OH^-]$, and vice versa. (Probs. 19.1, 19.30)

19.2 pH
 1. Define pH and interpret the meaning of the pH scale.
 *2. Given any one of the following, calculate the others: $[H^+]$, $[OH^-]$, pH. (Probs. 19.1, 19.2, 19.29, 19.30, 19.31)

19.3 Strong Acids and Bases
 1. Explain the meaning of "strong" as applied to acids and bases.
 *2. Identify the common strong acids and bases (Table 19.4) (Prob. 19.5)
 *3. Describe how to prepare a solution of a given $[H^+]$, $[OH^-]$, or pH from stock solutions of strong acids and bases. (Prob. 19.45)

19.4 Weak Acids
1. Explain the meaning of "weak" as applied to acids and bases.
*2. Classify a substance as a strong or weak acid when given its molecular formula. (Prob. 19.5)
3. Describe how to experimentally distinguish between strong and weak acids.
*4. Write the equilibrium constant expression for any weak acid.
*5. Calculate K_a when given concentration data for solutions of weak acids. (Probs. 19.2, 19.34)
*6. Calculate [H^+] and per cent dissociation when given the initial concentration and the value of K_a for a weak acid.
*7. Apply the "5% rule" and method of successive approximations to equilibrium problems involving the dissociation of weak acids. (Probs. 19.3, 19.37)

19.5 Weak Bases
*1. Classify a substance as a strong or weak base when given its molecular formula. (Prob. 19.5)
*2. Write the equilibrium constant expression for any weak base. (Prob. 19.37)
3. Explain the meaning of conjugate acid, conjugate base.
*4. Use the Law of Multiple Equilibria to calculate K_b for a weak base when given K_a for the conjugate acid. (Prob. 19.4)
*5. Determine K_b when given the pH, [H^+], or [OH^-] of a weak base at a known concentration. (Prob. 19.37)
*6. Calculate [OH^-], pH, and per cent dissociation when given the initial concentration and the value of K_b of a weak base. (Prob. 19.37)

19.6 Acid-Base Properties of Salt Solutions
*1. Given the formula of an ionic compound, classify the ions as acidic, basic, or neutral. (Probs. 19.6, 19.39)
*2. Predict whether a particular ion or ionic compound will form an acidic, basic, or neutral water solution; write net ionic equations to explain the acidity or basicity. (Probs. 19.6, 19.32, 19.39, 19.40)

19.7 General Theories of Acids and Bases
1. Define the terms acid and base according to the Brönsted-Lowry concept, and the Lewis concept.
*2. Identify the Brönsted acid and Brönsted base when given the equation for an acid-base reaction. (Probs. 19.7, 19.42)
*3. Identify the Lewis acid and the Lewis base when given the equation for an acid-base reaction. (Ex. 19.3)

SUMMARY

1. Aqueous Solutions and the pH Scale

Water dissociates to a very limited extent according to the equation: $H_2O(l) \rightleftharpoons H^+ (aq) + OH^-(aq)$; $K_W = [H^+][OH^-] = 1.0 \times 10^{-14}$. The K_W relationship holds for any solution at 25°C which has water as a solvent. Acids may be defined as substances which when added to water form solutions with $[H^+] > 10^{-7}$. Bases then are substances which when added to water form solutions with $[OH^-] > 10^{-7}$. In a neutral solution, $[H^+] = [OH^-]$. Acidity or basicity can be expressed in terms of pH, defined as $pH = -\log_{10}[H^+]$. It follows that an acidic solution has a pH less than 7, that of a basic solution is greater than 7, and a neutral solution has a pH equal to 7.

Exercise 19.1
For the following solutions, calculate:
1. $[H^+]$, if the $[OH^-] = 5.6 \times 10^{-4}$
2. $[OH^-]$, if the $[H^+] = 2.4 \times 10^{-10}$
3. pH, if the $[H^+] = 8.0 \times 10^{-3}$
4. pH, if the $[OH^-] = 7.6 \times 10^{-11}$
5. $[H^+]$, if the pH = 4.0
6. $[OH^-]$, if the pH = 9.20

Exercise 19.1 Answers
You will need to be able to work with logs and antilogs in some of these problems. Review Appendix 4.

1. $[H^+] = \dfrac{1.0 \times 10^{-14}}{[OH^-]} = \dfrac{1.0 \times 10^{-14}}{5.6 \times 10^{-4}} = 1.8 \times 10^{-11}$

2. $[OH^-] = \dfrac{1.0 \times 10^{-14}}{[H^+]} = \dfrac{1.0 \times 10^{-14}}{2.4 \times 10^{-10}} = 4.2 \times 10^{-5}$

3. $pH = -\log_{10}[H^+] = -\log_{10}(8.0 \times 10^{-3}) = -(0.9 - 3.0) = 2.1$

4. Use the K_W expression to find $[H^+]$.

   $[H^+] = \dfrac{1.0 \times 10^{-14}}{[OH^-]} = \dfrac{1.0 \times 10^{-14}}{7.6 \times 10^{-11}} = 1.3 \times 10^{-4}$

   $pH = -\log_{10}(1.3 \times 10^{-4}) = -(0.11 - 4.0) = 3.9$

5. $\log_{10}[H^+] = -4.0$; $[H^+] = 1.0 \times 10^{-4}$

6. $\log_{10} [H^+] = -9.20 = 0.80 - 10$; $[H^+] = 6.3 \times 10^{-10}$

$$[OH^-] = \frac{1.0 \times 10^{-14}}{[H^+]} = \frac{1.0 \times 10^{-14}}{6.3 \times 10^{-10}} = 1.6 \times 10^{-5}$$

2. Acid and Base Strengths

Strong acids and strong bases are substances that ionize completely, or almost completely, when added to water. The strengths of acids and bases are indicated by the magnitudes of their ionization constants, K_a and K_b, respectively. Anions derived from weak acids behave as weak bases in water solution. The equation $K_a K_b = K_w$, shows the relationship between the acid and base dissociation constants.

The quantitative relationships between K_a, K_b, and solution concentration are illustrated in examples in the text and by the following exercises.

Exercise 19.2
1. Calculate the volume of 2.0 M HNO_3 that must be used to prepare 840 ml of a solution that has a pH of 2.0.
2. A 0.0100 M solution of $HC_2H_3O_2$ has a pH = 3.38. Calculate the value of K_a for this weak acid.
3. Calculate the per cent dissociation of $HC_2H_3O_2$ based on the data in the above problem.
4. Calculate the $[H^+]$ in a 0.10 M solution of HOCl. The $K_a = 3.2 \times 10^{-8}$.
5. Hydrocyanic acid, HCN, has a $K_a = 4.0 \times 10^{-10}$. Calculate K_b for the cyanide ion.
6. Calculate the pH of a 1.0 M KCN solution.
7. Predict whether 0.1 M solutions of each of the following salts will be acidic, basic, or neutral.
   a. $KClO_4$ b. $FeCl_3$ c. K_2CO_3 d. NH_4Cl
8. Write a net ionic equation for the reaction, if any, for each of the salt solutions in problem 7.

Exercise 19.2 Answers
1. You need to recognize that HNO_3 is a strong acid and that pH = 2 = 1.0×10^{-2} = $[H^+]$. The number of moles of HNO_3 in 840 ml of 1.0×10^{-2} M solution is:

$$1.0 \times 10^{-2} \frac{\text{mole}}{\text{liter}} \times 0.840 \text{ liter} = 8.4 \times 10^{-3} \text{ moles}$$

The volume of 2.0 M stock solution required is:

8.4×10^{-3} moles $\times \dfrac{1 \text{ liter}}{2.0 \text{ moles}} = 4.2 \times 10^{-3}$ liter $= 4.2$ ml

The solution is prepared by adding water to 4.2 ml of the HNO_3 stock solution to give a total volume of 840 ml.

2. The equation for the dissociation is:

$$HC_2H_3O_2(aq) \rightleftharpoons H^+(aq) + C_2H_3O_2^-(aq); \quad K_a = \dfrac{[H^+][C_2H_3O_2^-]}{[HC_2H_3O_2]}$$

Note that $[H^+] = [C_2H_3O_2^-]$. Convert pH to $[H^+]$ to obtain the equilibrium concentrations of both of these ions. (Refer to the first step in the solution in Exercise 19.1, part (6) for this manipulation.) The $[H^+] = 4.2 \times 10^{-4} = [C_2H_3O_2^-]$. The equilibrium concentration of undissociated $HC_2H_3O_2$ will be 1.00×10^{-2} M $- 0.042 \times 10^{-2}$ M $= 0.96 \times 10^{-2}$ M $= 9.6 \times 10^{-3}$ M

Substituting: $K_a = \dfrac{(4.2 \times 10^{-4})(4.2 \times 10^{-4})}{9.6 \times 10^{-3}} = 1.8 \times 10^{-5}$

3. % dissociation $= \dfrac{[H^+] \times 100}{\text{orig conc } HC_2H_3O_2} = \dfrac{4.2 \times 10^{-4}}{1.00 \times 10^{-2}} \times 100 = 4.2\%$

4. The equation is: $HOCl(aq) \rightleftharpoons H^+(aq) + OCl^-(aq)$
Let $x = [H^+] = [OCl^-]$; $[HOCl] = 0.10 - x$

$$K_a = \dfrac{[H^+][OCl^-]}{[HOCl]} = 3.2 \times 10^{-8} = \dfrac{(x)(x)}{(0.10 - x)}$$

Since K_a has a small value, very little HOCl will dissociate and the equilibrium concentration must be nearly equal to the original concentration, 0.10M. Based on this approximation the expression is

$3.2 \times 10^{-8} = \dfrac{x^2}{0.10}$; $x^2 = 3.2 \times 10^{-9} = 32 \times 10^{-10}$; $x = 5.7 \times 10^{-5}$
$= [H^+]$

Apply the 5% rule to see if the above approximation is valid. In this case, 5.7×10^{-5} is much less than 5% of 0.10. The approximation is justified.

5. $K_b = \dfrac{K_w}{K_a} = \dfrac{1.0 \times 10^{-14}}{4.0 \times 10^{-10}} = 2.5 \times 10^{-5}$

6. The reaction is $CN^-(aq) + H_2O(l) \rightleftharpoons HCN(aq) + OH^-(aq)$
Let $x = [HCN] = [OH^-]$. Since K_b is relatively small (2.5×10^{-5}) and the initial concentration of CN^- is high, assume $[CN^-] \simeq 1.0$

$$K_b = \frac{[HCN][OH^-]}{[CN^-]} = 2.5 \times 10^{-5} = \frac{(x)(x)}{1.0};$$

$x^2 = 2.5 \times 10^{-5} = 25 \times 10^{-6}$; $x = 5.0 \times 10^{-3} = [OH^-]$

Checking the approximation, x is 0.5% of 1.0 and the assumption is justified.

$$[H^+] = \frac{K_w}{[OH^-]} = \frac{1.0 \times 10^{-14}}{5.0 \times 10^{-3}} = 2.0 \times 10^{-12}; pH = 11.7$$

7.

	salt	cation	anion	solution
a.	$KClO_4$	neutral	neutral	neutral
b.	$FeCl_3$	acidic	neutral	acidic
c.	K_2CO_3	neutral	basic	basic
d.	NH_4Cl	acidic	neutral	acidic

8. a. $KClO_4 \rightarrow$ no reaction
    b. $Fe(H_2O)_6^{3+}(aq) \rightarrow H^+(aq) + Fe(H_2O)_5(OH)^{2+}(aq)$
    c. $CO_3^{2-}(aq) + H_2O \rightarrow HCO_3^-(aq) + OH^-(aq)$
    d. $NH_4^+(aq) \rightarrow NH_3(aq) + H^+(aq)$

3. Acid-Base Concepts

The Brönsted-Lowry concept defines an acid as a proton donor, and a base as a proton acceptor. This is a general definition, in that it can be applied to solutions in which water is not the solvent. The Lewis concept considers an acid to be a species that can accept an electron pair; a base is the substance that can donate an electron pair. This is an even more general definition — it can be applied to reactions that do not involve a proton transfer.

Exercise 19.3
1. For each of the following reactions, indicate the Brönsted acids.

    a. $HSO_4^-(aq) + H_2O \rightleftharpoons SO_4^{2-}(aq) + H_3O^+(aq)$
    b. $NO_2^-(aq) + H_2O \rightleftharpoons HNO_2(aq) + OH^-(aq)$
    c. $HSO_4^-(aq) + HCO_3^-(aq) \rightleftharpoons SO_4^{2-}(aq) + H_2CO_3(aq)$

2. Use Table 19.5 to predict whether the equilibrium state of the system would favor products or reactants for each reaction in Problem 1.
3. Select the Lewis base in each of the following reactions.

   a. $Ag^+(aq) + 2NH_3(aq) \rightarrow Ag(NH_3)_2^+(aq)$

   b. $BCl_3 + NH_3 \rightarrow Cl_3BNH_3$

   c. $H^+(aq) + CN^-(aq) \rightleftharpoons HCN(aq)$

Exercise 19.3 Answers
1. a. HSO_4^-, H_3O^+
   b. H_2O, HNO_2
   c. HSO_4^-, H_2CO_3
2. a. Reactants. H_3O^+ is a stronger acid than HSO_4^-.
   b. Reactants. HNO_2 is a stronger acid than H_2O.
   c. Products. HSO_4^- is a stronger acid than H_2CO_3.
3. a. NH_3. The unshared pair of electrons on NH_3 is donated to the Ag^+.
   b. NH_3. The unshared pair of electrons is donated to an empty boron orbital.
   c. CN^-. An electron pair from CN^- is donated to the empty hydrogen orbital.

SUGGESTED ASSIGNMENT XIX

Text: Chapter 19

Problems: Numbers 19.1–19.7, 19.29–19.32, 19.34, 19.37, 19.39, 19.40, 19.42, 19.43, 19.45

Solutions to Assigned Problems Set XIX

19.1

To determine the $[OH^-]$, use the K_w expression in the form $[OH^-] = \dfrac{K_w}{[H^+]} = \dfrac{1.0 \times 10^{-14}}{2.0 \times 10^{-4}} = 5.0 \times 10^{-11}$

$pH = -\log_{10}[H^+] = -\log_{10}(2.0 \times 10^{-4}) = -(0.3 - 4.0) = 3.7$

19.2

The equation for the dissociation is

$HA(aq) \rightleftharpoons H^+(aq) + A^-(aq)$

We need to convert from pH to $[H^+]$.

$-\log_{10}[H^+] = 3.0$

$\log_{10}[H^+] = -3.0$. Taking antilogs, $[H^+] = 1.0 \times 10^{-3}$

$[A^-] = [H^+] = 1.0 \times 10^{-3}$

Since 1.0×10^{-3} mole of H^+ is formed, 1.0×10^{-3} moles of HA must dissociate. The equilibrium concentration of HA is $0.20\ M - 0.001\ M = 0.199\ M \approx 0.20\ M$

$K_a = \dfrac{[H^+][A^-]}{[HA]} = \dfrac{(1.0 \times 10^{-3})(1.0 \times 10^{-3})}{2.0 \times 10^{-1}} = 5.0 \times 10^{-6}$

19.3

We can represent the dissociation as

$HA(aq) \rightleftharpoons H^+(aq) + A^-(aq)$

Let $x = [H^+] = [A^-]$; $[HA] = 0.10 - x$

$$K_a = \frac{[H^+][A^-]}{[HA]} = 4.0 \times 10^{-5} = \frac{(x)(x)}{0.10 - x}$$

If we assume that x is small compared to 0.10,

$x^2 = 4.0 \times 10^{-6}$, $x = 2.0 \times 10^{-3} = [H^+]$

Since the value of x, 2.0×10^{-3} in this case, is less than 5% of 0.10, the approximation was valid.

19.4

The A^- is the conjugate base of the acid HA. As illustrated in Example 19.7 in the text, $K_b = \dfrac{K_w}{K_a}$

$$K_b = \frac{1.0 \times 10^{-14}}{4.0 \times 10^{-5}} = 2.5 \times 10^{-10}$$

19.5

Table 19.4 lists the important strong acids and bases. Table 19.5 lists the dissociation constants of some weak acids and bases.

a. NaOH is a strong base
b. HF is a weak acid
c. NH_4^+ is a weak acid
d. NH_3 is a weak base
e. F^- is a weak base
f. HI is a strong acid

19.6

We need to consider the ions from these soluble salts. Neutral cations are those derived from strong bases. Neutral anions are those derived from strong acids (Table 19.4). Cations derived from weak bases are acidic. Anions derived from weak acids are basic.

a. Consider $NaC_2H_3O_2$ as derived from NaOH (a strong base) and $HC_2H_3O_2$ (a weak acid). Thus the cation is neutral; the anion basic. The solution should be basic.

$$C_2H_3O_2^-(aq) + H_2O \rightleftharpoons HC_2H_3O_2(aq) + OH^-(aq)$$

b. $ZnCl_2$: acidic cation, neutral anion, solution acidic.

$$Zn(H_2O)_4^{2+}(aq) \rightarrow Zn(H_2O)_3(OH)^+(aq) + H^+(aq)$$

c. KNO_3: neutral cation, neutral anion, neutral solution, no reaction.

d. NH_4Br: acidic cation, neutral anion, acidic solution.

$$NH_4^+(aq) \rightleftharpoons NH_3(aq) + H^+(aq)$$

19.7

A Brönsted acid is a proton donor. A Brönsted base is a proton acceptor.

The H_3O^+ and H_2CO_3 are Brönsted acids. The HCO_3^- and H_2O act as Brönsted bases.

19.29

This problem is similar to Example 19.1 and Problem 19.1.

a. $pH = -\log_{10}[H^+] = -\log_{10}(1 \times 10^{-1}) = -(-1) = 1$

b. $pH = -\log_{10}[H^+] = -\log_{10}(1 \times 10^1) = -(1) = -1$

c. $pH = -\log_{10}[H^+] = -\log_{10}(7.0 \times 10^{-3}) = -(0.85 - 3.00) = 2.15$

d. $pH = -\log_{10}[H^+] = -\log_{10}(8.2 \times 10^{-9}) = -(0.91 - 9.00) = 8.09$

Solution (d) is basic.

19.30

a. $-\log_{10}[H^+] = 9.0; \log_{10}[H^+] = -9.0; [H^+] = 1.0 \times 10^{-9}$

$$[OH^-] = \frac{K_w}{[H^+]} = \frac{1.0 \times 10^{-14}}{1.0 \times 10^{-9}} = 1.0 \times 10^{-5}$$

b. $-\log_{10} [H^+] = 3.20; \log_{10} [H^+] = -3.20 = 0.80 - 4.00$
Taking antilogs, $[H^+] = 6.3 \times 10^{-4}$

$[OH^-] = \dfrac{1.0 \times 10^{-14}}{6.3 \times 10^{-4}} = 1.6 \times 10^{-11}$

c. $-\log_{10} [H^+] = -1.05; \log_{10} [H^+] = 1.05; [H^+] = 1.1 \times 10^1$

$[OH^-] = \dfrac{1.0 \times 10^{-14}}{1.1 \times 10^1} = 9.1 \times 10^{-16}$

d. $-\log_{10} [H^+] = 7.46; \log_{10} [H^+] = -7.46 = 0.54 - 8;$

$[H^+] = 3.47 \times 10^{-8}; [OH^-] = \dfrac{1.0 \times 10^{-14}}{3.5 \times 10^{-8}} = 2.9 \times 10^{-7}$

Solutios (b) and (c) are acidic.

19.31

a. This is a strong base. $[OH^-] = 8.0 \times 10^{-1}$

$[H^+] = \dfrac{K_w}{[OH^-]} = \dfrac{1.0 \times 10^{-14}}{8.0 \times 10^{-1}} = 1.3 \times 10^{-14}$

$pH = -\log_{10} (1.3 \times 10^{-14}) = -(0.11 - 14.00) = 13.9$

b. This is a strong acid. $[H^+] = 6.0 \times 10^{-1}$

$[OH^-] = \dfrac{1.0 \times 10^{-14}}{6.0 \times 10^{-1}} = 1.7 \times 10^{-14}$

$pH = -\log_{10} (6.0 \times 10^{-1}) = -(0.78 - 1.00) = 0.22$

c. Calculate the number of moles of the strong base and divide by the number of liters of final solution.

$0.0080 \text{ liter} \times \dfrac{6.0 \text{ mole}}{\text{liter}} \times \dfrac{1}{0.480 \text{ liter}} = 0.10 \text{ M} = [OH^-]$

$[H^+] = \dfrac{1.0 \times 10^{-14}}{1.0 \times 10^{-1}} = 1.0 \times 10^{-13}; pH = 13$

d. Determine the number of moles of the strong acid, and divide by the number of liters of solution.

$$75g \times \frac{1 \text{ mole}}{36.5g} \times \frac{1}{2.0 \text{ liters}} = 1.0M = [H^+]; [OH^-] = 1.0 \times 10^{-14}$$

$$pH = -\log_{10}[H^+] = 0$$

19.32

a. $H_3PO_4(aq) \rightleftharpoons H^+(aq) + H_2PO_4^{2-}(aq)$

b. $CO_2(g) + H_2O \rightleftharpoons H^+(aq) + HCO_3^-(aq)$

c. $HNO_2(aq) \rightleftharpoons H^+(aq) + NO_2^-(aq)$

d. $Al(H_2O)_6^{3+}(aq) \rightleftharpoons Al(H_2O)_5(OH)^{2+}(aq) + H^+(aq)$

The hydrated Al^{3+} is an acidic cation; the Cl^- a neutral anion.

19.34

$$H-\overset{\overset{O}{\|}}{C}-O-H$$ The H atom bonded to the oxygen is the acidic hydrogen.

$-\log_{10}[H^+] = 2.30; \log_{10}[H^+] = -2.30 = 0.70 - 3.00$

Taking antilogs, $[H^+] = 5.0 \times 10^{-3} = [HCOO^-]$

The number of moles of HCOOH dissolved in a liter of water is:

$$5.0g \times \frac{1 \text{ mole}}{46.0g} = 0.109 \text{ moles}$$

The $[HCOOH] = 0.109M - 0.005M = 0.104M$

(Remember that 0.0050 moles of the HCOOH dissociate to produce H^+ and $HCOO^-$.)

$HCOOH(aq) \rightleftharpoons H^+(aq) + HCOO^-(aq)$

$$K_a = \frac{[H^+][HCOO^-]}{[HCOOH]} = \frac{(5.0 \times 10^{-3})(5.0 \times 10^{-3})}{0.104} = 2.4 \times 10^{-4}$$

19.37

The Na^+ is a neutral cation. The Bz^- is basic anion. The equation is

$$Bz^-(aq) + H_2O \rightleftharpoons HBz(aq) + OH^-(aq)$$

a. From Table 19.5, the K_a of HBz is 6.6×10^{-5}

$$K_b = \frac{K_w}{K_a} = \frac{1.0 \times 10^{-14}}{6.6 \times 10^{-5}} = 1.5 \times 10^{-10}$$

b. Let $x = [HBz] = [OH^-]$; $[Bz^-] = 0.50 - x \approx 0.50$

$$K_b = \frac{[HBz][OH^-]}{[Bz^-]} = 1.5 \times 10^{-10} = \frac{(x)(x)}{0.50};$$

$x^2 = 7.5 \times 10^{-11} = 75 \times 10^{-12}$; $x = 8.7 \times 10^{-6} = [OH^-]$

Note that x is small when compared to 0.50. The approximation of the equilibrium concentration of Bz^- is justified.

$$[H^+] = \frac{K_w}{[OH^-]} = \frac{1.0 \times 10^{-14}}{8.7 \times 10^{-6}} = 1.1 \times 10^{-9}$$

$pH = -\log_{10}[H^+] = -\log_{10} 1.1 \times 10^{-9} = -(0.04 - 9.00) = 8.96$

19.39

This problem is similar to Problem 19.6.

a. $Al(NO_3)_3$: acidic cation, neutral anion, acidic solution.

b. HOCl: HOCl is a weak acid, acidic solution.

c. NaOCl: neutral cation, basic anion, basic solution.

d. NH_4NO_3: acidic cation, neutral anion, acidic solution.

e. Na_2CO_3: neutral cation, basic anion, basic solution.

f. $NaHSO_4$: neutral cation, acidic anion, acidic solution.

19.40

a. $Al(H_2O)_6^{3+}(aq) \rightleftharpoons Al(H_2O)_5(OH)^{2+}(aq) + H^+(aq)$

b. $HOCl(aq) \rightleftharpoons H^+(aq) + OCl^-(aq)$

c. $OCl^-(aq) + H_2O \rightleftharpoons HOCl(aq) + OH^-(aq)$

d. $NH_4^+(aq) \rightleftharpoons H^+(aq) + NH_3(aq)$

e. $CO_3^{2-}(aq) + H_2O \rightleftharpoons HCO_3^-(aq) + OH^-(aq)$

f. $HSO_4^-(aq) \rightleftharpoons H^+(aq) + SO_4^{2-}(aq)$

19.42

This is similar to Problem 19.7.

	Brönsted acids	Brönsted bases
a.	H_2O, HCN	CN^-, OH^-
b.	H_3O^+, H_2CO_3	HCO_3^-, H_2O
c.	$HC_2H_3O_2$, H_2S	HS^-, $C_2H_3O_2^-$

19.43

a. $K = \dfrac{[HCN][OH^-]}{[CN^-]} = K_b$; $CN^- = 2.5 \times 10^{-5}$

b. $K = \dfrac{[H_2CO_3]}{[HCO_3^-][H^+]} = \dfrac{1}{K_a \text{ of } H_2CO_3} = \dfrac{1}{4.2 \times 10^{-7}} = 2.4 \times 10^6$

c. $HC_2H_3O_2(aq) \rightleftharpoons H^+(aq) + C_2H_3O_2^-(aq)$; $K_I = K_a$ of $HC_2H_3O_2$

$H^+(aq) + HS^-(aq) \rightleftharpoons H_2S(aq)$; $K_{II} = \dfrac{1}{K_a \text{ of } H_2S}$

$HC_2H_3O_2(aq) + HS^-(aq) \rightleftharpoons C_2H_3O_2^-(aq) + H_2S(aq)$;
$K = (K_I)(K_{II})$

$$K = (K_a HC_2H_3O_2) \times \frac{1}{K_a \text{ of } H_2S} = \frac{1.8 \times 10^{-5}}{1 \times 10^{-7}} = 1.8 \times 10^2$$

19.45

A solution with a pH of 2.0 has a $[H^+] = 1.0 \times 10^{-2}$. For the HCl solution:

$$1.0 \times 10^{-2} \frac{\text{moles}}{\text{liter}} \times 6.0 \text{ liters} = 6.0 \times 10^{-2} \text{ moles}$$

Since HCl is a strong acid, 0.060 moles are required for six liters of solution.

For the chloroacetic acid solution: Since this is a relatively weak acid, solve the K_a expression for [HClA]. $HClA(aq) \rightleftharpoons H^+(aq) + ClA^-(aq)$

$$K_a = \frac{[H^+][ClA^-]}{[HClA]} = 1.4 \times 10^{-3} = \frac{(1.0 \times 10^{-2})(1.0 \times 10^{-2})}{[HClA]};$$

$[HClA] = 7.1 \times 10^{-2}$

The number of moles per liter of HClA at equilibrium is 7.1×10^{-2}. To produce 0.010 moles of H^+ per liter and have 0.071 moles per liter of HClA at equilibrium, we must start with: 0.071 moles + 0.010 moles = 0.081 moles per liter of HClA.

For six liters, $0.081 \frac{\text{moles}}{\text{liter}} \times 6.0 \text{ liter} = 0.49$ moles of HClA are required.

UNIT 20

Acid-Base Reactions

The goal of this unit is to consider different types of acid-base reactions and their applications to analytical chemistry, inorganic synthesis, and industrial processes.

INSTRUCTIONAL OBJECTIVES

20.1 Types of Acid-Base Reactions
 *1. Write a net ionic equation for any acid-base reaction. (Probs. 20.1, 20.29)
 2. Realize that $K = 1/K_w$ for a strong acid–strong base reaction.
 *3. Use the Law of Multiple Equilibria to calculate K for an acid-base reaction.
    *a. Calculate K for a reaction between a weak acid and a strong base when given K_b for the conjugate base. (Probs. 20.2b, 20.30)
    *b. Calculate K for a reaction between a strong acid and weak base when given K_a for the conjugate acid. (Probs. 20.2c, 20.30)
    *c. Use the derived equilibrium constants to calculate the concentration of any species in solution. (Prob. 20.47)
    *d. Calculate K for a reaction involving acid-base equilibria and a solubility product. Use K to determine whether a precipitate will form and whether a precipitate will be soluble in a solution of a given pH. (Probs. 20.31, 20.47)

20.2 Acid-Base Titrations
 *1. Calculate the concentration of a solution of an acid or base, given titration data. (Probs. 20.3, 20.34)
 *2. Calculate the gram equivalent weight of an acid, given titration data. (Prob. 20.35)
 *3. Relate the normality of a reagent to its molarity, or its gram

equivalent weight to its gram molecular weight, given the equation for an acid-base reaction.

*4. Use equilibrium concepts to explain how an acid-base indicator works; calculate the ratio $[A^-]/[HA]$ to determine the color of an indicator, given K_a and the concentration of H^+. (Probs. 20.4, 20.37)

5. Distinguish between end point and equivalence point.

*6. Calculate the pH at the equivalence for a titration, given the formulas and concentrations of the acid and base and the K_a and K_b. (Prob. 20.39)

*7. Select an appropriate indicator for a particular titration, given K_a values of several indicators. (Prob. 20.37)

*8. Draw and/or interpret a titration curve when given titration data. (Prob. 20.39)

20.3 Buffers

1. Explain the principle of buffer action.

*2. Given the composition of a buffered solution, calculate the pH of the solution before and after the addition of a known amount of a strong acid or base. (Probs. 20.5, 20.33, 20.42)

*3. Calculate K_a, given or having calculated the ratio $[A^-]/[HA]$ and the $[H^+]$. (Prob. 20.43)

20.4–20.6 Applications of Acid-Base Reactions

*1. Explain how to use acid-base reactions to prepare salts and certain volatile acids and bases. (Prob. 20.44)

*2. Explain how to use acid-base reactions to separate and/or identify ions. (Probs. 20.6, 20.45)

3. Describe the principles upon which the Solvay Process is based.

SUMMARY

1. Types of Acid-Base Reactions

When acids and bases react, the reaction usually proceeds to the right, yielding relatively large concentrations of products. The equilibrium constants for such reactions are usually large, but dependent upon K_w, K_a, and K_b (and K_{sp} if one of the reactants is a solid). The Law of Multiple Equilibria may be applied to calculate the values of K for these reactions.

A buffer solution is a solution that resists changes in pH when moderate quantities of acid or base are added. One type of buffer solution is formed by mixing a conjugate acid-base pair, HA and A^-; where HA is a weak acid and A⁻ is the anion of the weak acid. A buffer solution can also be prepared by mixing a weak base with one of its salts. The $[H^+]$ for a buffer solution may be derived from the equilibrium constant expression in the form

$$[H^+] = K_a \frac{[HA]}{[A^-]}$$

In a buffered solution, the weak acid reacts with OH^- that is added; while the anion of the weak acid reacts with any H^+ that may be introduced. In neither case does the pH change appreciably as long as moderate amounts of acid or base are added.

The following exercises illustrate an application of the Law of Multiple Equilibria and the preparation of a buffered solution to a given pH.

Exercise 20.1

1. Calculate the $[OH^-]$ in a solution prepared by mixing 325 ml of 0.100 M HCl solution with 175 ml of 0.100 M $Ba(OH)_2$ solution.
2. Calculate the $[OH^-]$ at equilibrium in a solution that is prepared by mixing 5 ml of 0.40 M HCN with 5 ml of 0.40 NaOH. K_b CN^- = 2.5 × 10⁻⁵
3. What molar ratio of $HCHO_2/NaCHO_2$ would be required to prepare a buffer solution of pH = 4? The K_a for formic acid = 2 × 10⁻⁴.

Exercise 20.1 Answers

1. We recognize that this is a reaction between a strong acid and a strong base. The net ionic equation is

$$H^+(aq) + OH^-(aq) \rightarrow H_2O$$

We need to calculate the number of moles of H^+ and OH^- present and determine which, if either, is in excess.

Number of moles H^+: $0.325 \text{ liter} \times 0.100 \frac{\text{mole } H^+}{\text{liter}} = 0.0325$ mole H^+

Number of moles OH^-: $0.175 \text{ liter} \times 0.100 \frac{\text{mole } Ba(OH)_2}{\text{liter}}$

$\times \frac{2 \text{ mole } OH^-}{\text{mole } Ba(OH)_2} = 0.0350$ mole OH^-

The 0.0325 mole H^+ will react with 0.0325 mole OH^- forming

0.0325 mole H_2O. The excess 0.0025 mole OH^- is in 500 ml of solution.

$$[OH^-] = \frac{2.5 \times 10^{-3} \text{ mole}}{0.500 \text{ liter}} = 5.0 \times 10^{-3}$$

2. This is a reaction between a weak acid and a strong base.

The equation is: $HCN(aq) + OH^-(aq) \rightleftharpoons H_2O + CN^-(aq)$

For the reaction: $K = \dfrac{1}{K_b} = \dfrac{1}{2.5 \times 10^{-5}} = 4.0 \times 10^4$

Original number of moles HCN: $0.005 \text{ liter} \times 0.40 \dfrac{\text{mole HCN}}{\text{liter}} =$ 0.002 mole HCN

Original number of moles OH^-: $0.005 \text{ liter} \times 0.40 \dfrac{\text{mole } OH^-}{\text{liter}} =$ 0.002 mole OH^-

If the reaction proceeds to completion, 0.002 mole of OH^- react with 0.002 mole HCN to produce 0.002 mole H_2O and 0.002 mole CN^-.

The 0.002 mole CN^- is in 0.010 liter of solution.

$$[CN^-] = \frac{0.002 \text{ mole}}{0.010 \text{ liter}} = 0.2 \text{ M}$$

Let $x = [HCN] = [OH^-]$; then $0.2 - x = [CN^-]$. Although K for the reaction is not extremely large, 4.0×10^4, assume that the equilibrium concentration of CN^- is essentially 0.2 M. (Note that the $[CN^-]$ is known to only one significant figure.)

$$K = \frac{[CN^-]}{[HCN][OH^-]} = 4.0 \times 10^4 = \frac{0.2}{(x)(x)} \;;\; x^2 = \frac{0.2}{4.0 \times 10^4} =$$
5×10^{-6}; $x = 2 \times 10^{-3} = [OH^-]$

Using the 5% rule, $\dfrac{0.002}{0.2} \times 100 = 1\%$, the assumption regarding the $[CN^-]$ is correct.

3. $pH = 4$; $1 \times 10^{-4} = [H^+]$

The equilibrium expression $K_a = \dfrac{[H^+][CHO_2^-]}{[HCHO_2]}$ may be rewritten in the form indicated and the ratio calculated.

$$\dfrac{[HCHO_2]}{[CHO_2^-]} = \dfrac{[H^+]}{K_a} = \dfrac{1 \times 10^{-4}}{2 \times 10^{-4}} = 0.5$$

There must be one mole of formic acid for every two moles of sodium formate.

2. Acid-Base Titrations

The volume of solution required to react with a given quantity of another substance is often determined by a process called titration. In an acid-base titration, the reaction may be monitored by following the pH of the solution as a function of the volume of reagent that is added. Titration curves are shown in Figures 20.1, 20.2, 20.3, and 20.4 in the text. Note that the pH at the equivalence point for a given titration depends on the relative strengths of the acid and base involved.

Indicators may be used to determine the pH of a solution and thus the end point in a titration. Acid-base indicators are weak acids (or weak bases) in which the undissociated molecule has a color that is different from that of its conjugate base. For a particular titration, an indicator is selected so that the K_a of the indicator is equal to the hydrogen ion concentration at the equivalence point. The choice of an indicator is especially important in the titration of a weak acid with a strong base and for the titration of a weak base with a strong acid. For these titrations, the equivalence point is not at pH=7. Further, the rate of change of pH at the equivalence point for these kinds of titrations is quite slow compared to that for a strong acid–strong base titration.

The concepts of normality and gram-equivalent weight are sometimes used in calculations involving titration data. Keep in mind that the normality of a given reagent depends on the reaction in which it participates.

Exercise 20.2
1. Refer to Exercise 20.1(2) in which the [OH$^-$] was calculated. Assume that this was a titration reaction and calculate the pH at the equivalence point. Figure 20.3 in the text shows a titration curve for an HCN-NaOH reaction. The shape of the curve for the reaction in 20.1(2) would be very similar. Explain why the concentration of an HCN solution can not be determined by using an acid-base indicator in a direct titration.
2. The K_a value for a certain weak acid is 1×10^{-6}. The acid form is

red; the conjugate base is yellow. What color would the indicator have in a solution of pH = 5? 6? 7?

3. It requires 65.00 ml of 0.0500 M HCl solution to titrate 35.00 ml of a $Sr(OH)_2$ solution.

   Calculate the molarity of the $Sr(OH)_2$ solution.

   What is the normality of the $Sr(OH)_2$ solution?

Exercise 20.2   Answers

1. $[OH^-] = 2 \times 10^{-3}$; $[H^+] = \dfrac{K_w}{[OH^-]} = \dfrac{1.0 \times 10^{-14}}{2 \times 10^{-3}} = 5 \times 10^{-12}$

   pH = 11.3. The curve in Figure 20.3 indicates that there is not a sharp change in the pH at the equivalence point. Since the pH changes so gradually, an indicator would not show a rapid change in color. The lack of steepness in the curve is due to the fact that K for the reaction is relatively small.

2. $\dfrac{[In^-]}{[HIn]} = \dfrac{K_a}{[H^+]} = \dfrac{1 \times 10^{-6}}{[H^+]}$; pH = 5, $[H^+] = 1 \times 10^{-5}$

   pH = 6, $[H^+] = 1 \times 10^{-6}$

   pH = 7, $[H^+] = 1 \times 10^{-7}$

   $\dfrac{[In^-]}{[HIn]} = \dfrac{1 \times 10^{-6}}{1 \times 10^{-5}} = 0.10$; Molecular acid predominates; the color is red.

   $\dfrac{[In^-]}{[HIn]} = \dfrac{1 \times 10^{-6}}{1 \times 10^{-6}} = 1$; The concentrations of the acid form and base are equal; the color is orange.

   $\dfrac{[In^-]}{[HIn]} = \dfrac{1 \times 10^{-6}}{1 \times 10^{-7}} = 10$; The base predominates; the color is yellow.

3. This is a reaction between a strong acid and a strong base,

   $H^+(aq) + OH^-(aq) \rightarrow H_2O$

   number moles H^+ = number moles OH^- = (0.0500)(0.0650) = 3.25×10^{-3}

   number moles $Sr(OH)_2 = (3.25 \times 10^{-3})/2 = 1.62 \times 10^{-3}$

   $M\ Sr(OH)_2 = \dfrac{1.62 \times 10^{-3}\ \text{mole}}{3.50 \times 10^{-2}\ \text{liter}} = 0.0464\ M$

The normality of the $Sr(OH)_2$ (2 moles OH^- per mole base) is twice the molarity, or 0.0928 N.

3. Applications of Acid-Base Reactions

The principles of acid-base reactions are used in qualitative and quantitative chemical analyses as well as the synthesis of inorganic compounds. An industrial application is the production of $NaHCO_3$ and Na_2CO_3 by the Solvay Process. Applications to analytical methods are illustrated in the following exercises.

Exercise 20.3
1. It requires 32.48 ml of a 0.200 M solution of HCl to react with all the Na_2CO_3 in a 0.808g sample of a mixture of Na_2CO_3 and NaCl. Calculate the percentage of Na_2CO_3 in the sample.
2. A solution containing 1×10^{-6} mole per liter of $Hg^{2+}(aq)$ is saturated (0.10 M) with H_2S. At what concentration of H^+ would HgS begin to precipitate? The K_{sp} of $HgS = 1 \times 10^{-50}$;

   K_a for the reaction $H_2S(aq) \rightleftharpoons 2H^+(aq) + S^{2-}(aq)$ is 1×10^{-22}.

Exercise 20.3   Answers
1. The reaction is $2H^+(aq) + CO_3^{2-}(aq) \rightarrow H_2CO_3$

   number moles $H^+ = (0.200)(0.03248) = 6.50 \times 10^{-3}$

   number moles $Na_2CO_3 = (6.50 \times 10^{-3})/2 = 3.25 \times 10^{-3}$

   number grams $Na_2CO_3 = 3.25 \times 10^{-3}$ mole $\times \dfrac{106g}{1 \text{ mole}} = 0.344$ g

   % $Na_2CO_3 = \dfrac{0.344}{0.808} \times 100 = 42.6\%$

2. The reaction in this multiple equilibrium system may be considered to be: $HgS(s) + 2H^+(aq) \rightleftharpoons Hg^{2+}(aq) + H_2S(aq)$

   $K = \dfrac{[H_2S][Hg^{2+}]}{[H^+]^2} = \dfrac{K_{sp}}{K_a} = \dfrac{1 \times 10^{-50}}{1 \times 10^{-22}} = 1 \times 10^{-28}$

   $[H^+]^2 = \dfrac{[H_2S][Hg^{2+}]}{K} = \dfrac{(0.1)(1 \times 10^{-6})}{1 \times 10^{-28}} = 1 \times 10^{21} = 10 \times 10^{20}$

   $[H^+] = 3 \times 10^{10}$. At concentrations of H^+ less than 3×10^{10} the HgS will precipiate from a 1×10^{-6} M Hg^{2+} solution saturated with H_2S.

234 • UNIT 20–ACID-BASE REACTIONS

SUGGESTED ASSIGNMENT XX

Text: Chapter 20

Problems: Numbers   20.1–20.6,   20.29–20.31,   20.33–20.35,   20.37, 20.39, 20.42–20.45, 20.47

Solutions to Assigned Problems Set XX

20.1

a. Recognize that this is a reaction between a strong base and a strong acid.

$$H^+(aq) + OH^-(aq) \rightleftharpoons H_2O$$

b. This is a reaction between a strong base and a weak acid. Since the acid is weak, the molecular species is predominant.

$$HC_2H_3O_2(aq) + OH^-(aq) \rightleftharpoons H_2O + C_2H_3O_2^-(aq)$$

c. $H^+(aq) + NH_3(aq) \rightleftharpoons NH_4^+(aq)$

d. $ZnCO_3(s) + 2H^+(aq) \rightleftharpoons Zn^{2+}(aq) + H_2CO_3(aq)$

20.2

a. This reaction is the reverse of the dissociation of water.

$$K = \frac{1}{K_w} = \frac{1}{1.0 \times 10^{-14}} = 1.0 \times 10^{14}$$

Note that the above equation and this value of K are the same for all strong acid–strong base reactions.

b. The ionic equation in (b) is the reverse of the reaction of a basic anion with water.

$$K = \frac{1}{K_b \text{ of } C_2H_3O_2^-} = \frac{1}{5.6 \times 10^{-10}} = 1.8 \times 10^9$$

c. The ionic equation in (c) is the reverse of the ionization of the weak acid NH_4^+.

$$K = \frac{1}{K_a \text{ of } NH_4^+} = \frac{1}{5.6 \times 10^{-10}} = 1.8 \times 10^9$$

20.3

The ionic equation for the reaction is

$$H^+(aq) + OH^-(aq) \rightarrow H_2O$$

One mole of HCl reacts with one mole of NaOH. Using conversion factors:

$$0.050 \text{ liter NaOH} \times \frac{0.30 \text{ mole}}{\text{liter NaOH}} \times \frac{1 \text{ mole HCl}}{1 \text{ mole NaOH}} \times \frac{1}{0.015 \text{ liter HCl}} = \frac{1.0 \text{ mole HCl}}{\text{liter HCl}} = 1.0 \text{ M}$$

20.4

Methyl red is a weak acid. The ionic equation is $HIn(aq) \rightleftharpoons H^+(aq) + In^-(aq)$ (The position of the dissociation equilibrium is determined by the $[H^+]$.)

$$K_a = \frac{[H^+][In^-]}{[HIn]} \; ; \; \frac{[In^-]}{[HIn]} = \frac{K_a}{[H^+]} = \frac{1 \times 10^{-5}}{[H^+]}$$

When the pH = 3, the $[H^+] = 1 \times 10^{-3}$

$\frac{[In^-]}{[HIn]} = \frac{1 \times 10^{-5}}{1 \times 10^{-3}} = 1 \times 10^{-2}$ There is one In^- ion (yellow) for every 100 HIn molecules (red). The color is red. When the pH = 5, the $[H^+] = 1 \times 10^{-5}$

$\frac{[In^-]}{[HIn]} = \frac{1 \times 10^{-5}}{1 \times 10^{-5}} = 1$ There are equal amounts of In^- ion and molecular HIn. The solution appears orange (equal amounts of red and yellow).

When the pH = 7, the $[H^+] = 1 \times 10^{-7}$

$\frac{[In^-]}{[HIn]} = \frac{1 \times 10^{-5}}{1 \times 10^{-7}} = 1 \times 10^2$ There are 100 In^- ions for every HIn molecule. The color is yellow.

20.5

The pH of a buffer solution is established by the equilibrium between the NH_4^+ ion and NH_3.

$NH_4^+(aq) \rightleftharpoons H^+(aq) + NH_3(aq)$

$K_a = \dfrac{[H^+][NH_3]}{[NH_4^+]}$; $[H^+] = 5.6 \times 10^{-10} \dfrac{[NH_4^+]}{[NH_3]} = 5.6 \times 10^{-10} \times \dfrac{0.20}{0.40}$;

$[H^+] = 2.8 \times 10^{-10}$

$pH = -\log_{10}(2.8 \times 10^{-10}) = -(0.45 - 10.00) = 9.55$

20.6

a. Add excess strong acid and warm to produce $CO_2(g)$.

$2H^+(aq) + CO_3^{2-}(aq) \rightarrow H_2O + CO_2(g)$

We could also add Ba^{2+} to produce the white precipitate $BaCO_3$, or bubble the evolved CO_2 into a solution of $Ba(OH)_2$ to produce $BaCO_3$.

b. Add strong acid to produce $H_2S(g)$

$2H^+(aq) + S^{2-}(aq) \rightarrow H_2S(g)$

We could also add a solution of a metal ion to produce an insoluble metal sulfide.

c. Add a strong base to produce $NH_3(g)$ and heat.

$NH_4^+(aq) + OH^-(aq) \rightarrow NH_3(g) + H_2O$

Test the evolved gas with moist litmus for a basic reaction.

20.29

a. $HF(aq) + OH^-(aq) \rightleftharpoons H_2O + F^-(aq)$

b. $H^+(aq) + C_2H_3O_2^-(aq) \rightleftharpoons HC_2H_3O_2(aq)$

c. $NH_4^+(aq) + OH^-(aq) \rightleftharpoons NH_3(aq) + H_2O$

d. no reaction

e. $H^+(aq) + OH^-(aq) \rightarrow H_2O$

f. no reaction

20.30

Refer to Table 19.5 for the appropriate constants.

a. $K = 1/K_b$ for $CO_3^{2-} = \dfrac{1}{2.1 \times 10^{-4}} = 4.8 \times 10^3$

b. $K = 1/K_a$ for $H_2CO_3 = \dfrac{1}{4.2 \times 10^{-7}} = 2.4 \times 10^6$

c. Apply the Law of Multiple Equilibria.

$H^+(aq) + CO_3^{2-}(aq) \rightleftharpoons HCO_3^-(aq);\quad K_I = 1/K_a$ for HCO_3^-

$\underline{H^+(aq) + HCO_3^-(aq) \rightleftharpoons H_2CO_3(aq);\quad K_{II} = 1/K_a \text{ for } H_2CO_3}$

$2H^+(aq) + CO_3^{2-}(aq) \rightleftharpoons H_2CO_3(aq);\quad K = K_I \times K_{II}$

$K = \dfrac{1}{K_a HCO_3} \times \dfrac{1}{K_a H_2CO_3} = \left(\dfrac{1}{4.8 \times 10^{-11}}\right)\left(\dfrac{1}{4.2 \times 10^{-7}}\right) = \dfrac{1}{2.0 \times 10^{-17}} = 5.0 \times 10^{16}$

20.31

Apply the Law of Multiple Equilibria

a. $MnS(s) \rightleftharpoons Mn^{2+}(aq) + S^{2-}(aq);\qquad K_I = K_{sp}MnS$

$H^+(aq) + S^{2-}(aq) \rightleftharpoons HS^-(aq);\qquad K_{II} = 1/K_a HS^-$

$\underline{H^+(aq) + HS^-(aq) \rightleftharpoons H_2S(aq);\qquad K_{III} = 1/K_a H_2S}$

$MnS(s) + 2H^+(aq) \rightleftharpoons Mn^{2+}(aq) + H_2S(aq);\ K = K_I \times K_{II} \times K_{III}$

$K = \dfrac{K_{sp}MnS}{(K_a HS^-)(K_a H_2S)} = \dfrac{1 \times 10^{-13}}{(1 \times 10^{-15})(1 \times 10^{-7})} = 1 \times 10^9$

b. $Al(OH)_3(s) \rightleftharpoons Al^{3+}(aq) + 3OH^-(aq);$ $\quad K_I = K_{sp} Al(OH)_3$

$\underline{3H^+(aq) + 3OH^-(aq) \rightleftharpoons 3H_2O \qquad ;} \quad K_{II} = 1/(K_w)^3$

$Al(OH)_3(s) + 3H^+(aq) \rightleftharpoons Al^{3+}(aq) + 3H_2O; \quad K = K_I \times K_{II}$

$$K = \frac{K_{sp}Al(OH)_3}{(K_w)^3} = \frac{5 \times 10^{-33}}{(1 \times 10^{-14})^3} = \frac{5 \times 10^{-33}}{1 \times 10^{-42}} = 5 \times 10^9$$

Note that for K_{II}, we are adding the equation involving H_2O three times. The Law of Multiple Equilibria states that when we add equations, the values of K are multiplied. In this case, $1/K_w$ becomes $1/(K_w)^3$

20.33

The equation for the reaction is

$HAc(aq) + OH^-(aq) \rightarrow H_2O + Ac^-(aq)$

The number of moles of $HC_2H_3O_2$ available for reaction is

$\dfrac{1.0 \text{ mole}}{\text{liter}} \times 0.200 \text{ liter} = 0.200 \text{ mole}$

0.10 mole of NaOH is available for reaction.

$(0.200 - 0.10)$ mole $= 0.10$ mole $HC_2H_3O_2$ remaining $= 0.10$ mole $C_2H_3O_2^-$ formed.

$[HC_2H_3O_2] = \dfrac{0.10 \text{ mole}}{0.20 \text{ liter}} = 0.50 \text{ M} = [C_2H_3O_2^-]$

$K_a = \dfrac{[H^+][C_2H_3O_2^-]}{[HC_2H_3O_2]} ; \quad [H^+] = K_a \times \dfrac{[HC_2H_3O_2]}{[C_2H_3O_2^-]} = 1.8 \times 10^{-5} \times \dfrac{0.50}{0.50} = 1.8 \times 10^{-5}$

We recognize that this is a buffer solution since both $HC_2H_3O_2$ and $C_2H_3O_2^-$ are present.

20.34

This is similar to Problem 20.3 except that we must recognize that each mole of H_2SO_4 contains $2H^+$. Using conversion factors:

$$0.2137 \frac{\text{mole NaOH}}{\text{liter NaOH}} \times 0.02366 \text{ liter NaOH} \times \frac{1 \text{ mole } H_2SO_4}{2 \text{ moles NaOH}} \times$$

$$\frac{1}{0.02204 \text{ liter}} = 0.1147 \frac{\text{mole } H_2SO_4}{\text{liter}} = 0.1147 \text{ M}$$

Note that the answer contains four significant figures.

20.35

$$0.04365 \text{ liter} \times 0.527 \frac{\text{mole}}{\text{liter}} = 0.0230 \text{ mole NaOH} = \text{number of GEW}$$

NaOH = number of GEW acid

$$1.48\text{g} \times \frac{1}{0.0230 \text{ GEW}} = 64.3\text{g/GEW}$$

20.37

This is similar to Problem 20.4.

$$HIn(aq) \rightleftharpoons H^+(aq) + In^-(aq); \quad \frac{[In^-]}{[HIn]} = \frac{K_a}{[H^+]} = \frac{1 \times 10^{-5}}{[H^+]}$$

yellow $\qquad\qquad\qquad$ blue

Summarizing:

pH	1	3	4	5	7	10
$[H^+]$	1×10^{-1}	1×10^{-3}	1×10^{-4}	1×10^{-5}	1×10^{-7}	1×10^{-10}
$\frac{[In^-]}{[HIn]}$	1×10^{-4}	1×10^{-2}	1×10^{-1}	1	1×10^{2}	1×10^{5}
color	yellow	yellow	yellow	green	blue	blue

Bromcresol green changes color from yellow to blue as the pH goes from 4 to 6 and will appear green at the end point, pH 5 (when there are equal amounts of yellow and blue).

Bromcresol green would be best for the titration of NH_3 and HNO_3 since the pH at the end point is on the acidic side. In an NaOH + HCl titration, the pH changes very rapidly near the end point of pH 7. Bromcresol green would be satisfactory for this strong acid–strong base titration. However, bromcresol green would not be a good choice for the NaOH–NH$_4$Cl titration, since the end point is on the basic side.

20.39

The equation for the reaction that occurs during this titration is

$$H^+(aq) + OH^-(aq) \rightleftharpoons H_2O; K = 1.0 \times 10^{14}$$

a. $[OH^-] = 2.000 \times 10^{-1}$; $[H^+] = \dfrac{1.0 \times 10^{-14}}{2.000 \times 10^{-1}} = 5.0 \times 10^{-14}$

$pH = -\log_{10}(5.0 \times 10^{-14}) = -(0.70 - 14.000) = 13.30$

In the following problems, remember to add the volumes of HCl that are added to the 50.00 ml of NaOH before determining the concentrations.

b. There are 25.00 ml of unreacted NaOH in a total volume of 75.00 ml.

$[OH^-] = \dfrac{0.02500 \text{ liter} \times 0.2000 \text{ mole/liter}}{0.07500 \text{ liter}} = 6.667 \times 10^{-2}$

$[H^+] = \dfrac{1.0 \times 10^{-14}}{6.667 \times 10^{-2}} = 1.5 \times 10^{-13}$;

$pH = -\log_{10}(1.5 \times 10^{-13}) = 12.82$

c. There are $(50.00 - 49.99)$ ml $= 0.01$ ml $= 1 \times 10^{-5}$ liter of unreacted 0.2000 M NaOH in a total volume of about 100 ml of solution.

$[OH^-] = \dfrac{1 \times 10^{-5} \text{ liter} \times 0.2000 \text{ mole/liter}}{0.100 \text{ liter}} = 2.0 \times 10^{-5}$

$[H^+] = \dfrac{1.0 \times 10^{-14}}{2.0 \times 10^{-5}} = 5.0 \times 10^{-10}$; $pH = 9.30$

d. The HCl reacts with all of the NaOH to produce a solution of NaCl. The pH is that of pure H$_2$O, 7.00. This is the equivalence point in the titration. The pH has changed from 9.3 to 7 on the addition of only 0.01 ml of acid.

e. The H^+ is now in excess.

$(50.10 - 50.00)\text{ml} = 0.10\text{ ml} = 1.0 \times 10^{-4}$ liter

The total value is $100.10\text{ ml} \approx 1.00 \times 10^{-1}$ liter

$$[H^+] = \frac{1.0 \times 10^{-4}\text{ liter} \times 0.2000\text{ mole/liter}}{1.00 \times 10^{-1}\text{ liter}} = 2.0 \times 10^{-4}$$

$\text{pH} = -\log_{10}(2.0 \times 10^{-4}) = 3.70$

f. Here we have 50.00 ml of 0.2000 M HCl in a total volume of 150.00 ml of solution.

$$[H^+] = \frac{0.05000\text{ liter} \times 0.2000\text{ mole/liter}}{0.15000\text{ liter}} = 6.67 \times 10^{-2}$$

$\text{pH} = 1.18$

20.42

a. $\text{HLac(aq)} \rightleftharpoons H^+(aq) + \text{Lac}^-(aq);\qquad K_I = K_a = 8.4 \times 10^{-4}$

$H^+(aq) + HCO_3^-(aq) \rightleftharpoons H_2CO_3(aq);\quad K_{II} = 1/K_a = 1/(4.2 \times 10^{-7})$

$\text{HLac(aq)} + HCO_3^-(aq) \rightleftharpoons H_2CO_3(aq) + \text{Lac}^-(aq)$

$K = (K_I)(K_{II}) = (8.4 \times 10^{-4})/(4.2 \times 10^{-7}) = 2.0 \times 10^3$

b. $H_2CO_3(aq) \rightleftharpoons H^+(aq) + HCO_3^-(aq); K_a = 4.2 \times 10^{-7}$

$$[H^+] = K_a \frac{[H_2CO_3]}{[HCO_3^-]} = (4.2 \times 10^{-7})\frac{(0.0014)}{(0.027)} = 2.2 \times 10^{-8}$$

$\text{pH} = -\log_{10}(2.2 \times 10^{-8}) = 7.66$

c. Since K for the reaction is large, all of the HLac reacts to increase the concentration of H_2CO_3 and decrease the concentration of HCO_3^-.

$[H_2CO_3] = 0.0014 + 0.005 = 0.0064$

$[HCO_3^-] = 0.027 - 0.005 = 0.022$

$[H^+] = (4.2 \times 10^{-7}) \dfrac{(0.0064)}{(0.022)} = 1.22 \times 10^{-7}$

$pH = -\log_{10} (1.22 \times 10^{-7}) = 6.91$

20.43

The reaction for one-third of the original solution is

$HA(aq) + OH^-(aq) \rightleftharpoons H_2O + A^-(aq)$

When the titrated solution is added to the remaining original solution, the [HA] is two times the concentration of $[A^-]$. Therefore,

$[A^-] = \dfrac{[HA]}{2}$ and when pH = 6.4, $[H^+] = 4.0 \times 10^{-7}$

$K_a = \dfrac{[H^+][A^-]}{[HA]} = \dfrac{[H^+]}{[HA]} \times \dfrac{[HA]}{2} = \dfrac{4.0 \times 10^{-7}}{2} = 2.0 \times 10^{-7}$

20.44

a. Add a strong base such as NaOH and heat. Gaseous NH_3 is evolved.

b. Add NaOH to form $Cd(OH)_2$ precipitate, and filter. Treat the solid $Cd(OH)_2$ with an equivalent amount of H_2SO_4. Evaporate the solution to form $CdSO_4$ crystals.

c. Add HCl and evaporate.

d. Bubble CO_2 through the NaOH solution until the solution is saturated with CO_2. Evaporate the resulting solution to obtain $NaHCO_3$.

20.45

Add NaOH to the mixture and heat. The NH_4^+ is converted to $NH_3(g)$. The evolved NH_3 will turn moist litmus blue. The equation for the reaction is

$NH_4^+(aq) + OH^-(aq) \rightarrow NH_3(g) + H_2O$

Add excess HNO_3 to the remaining solution. Pass the evolved gas into a solution of $Ba(OH)_2$. The formation of a white precipitate indicates the presence of CO_2.

$CO_3^{2-}(aq) + 2H^+(aq) \rightarrow CO_2(g) + H_2O$

$CO_2(g) + Ba^{2+}(aq) + 2OH^-(aq) \rightarrow BaCO_3(s) + H_2O$

Add $Ba(NO_3)_2$ to the remaining solution to precipitate the SO_4^{2-}.

$Ba^{2+}(aq) + SO_4^{2-}(aq) \rightarrow BaSO_4(s)$

The presence of the Cl^- remaining in solution can be identified by adding $AgNO_3$.

$Ag^+(aq) + Cl^-(aq) \rightarrow AgCl(s)$

20.47

The least soluble sulfide (the one with the smallest K_{sp} value) will precipitate first. The order is $ZnS(K_{sp} = 1 \times 10^{-23})$, then both CoS and NiS (both have a $K_{sp} = 1 \times 10^{-21}$), then FeS ($K_{sp} = 1 \times 10^{-18}$), and finally MnS ($K_{sp} = 1 \times 10^{-13}$)

The equation for the reaction is

$FeS(s) + 2H^+(aq) \rightleftharpoons Fe^{2+}(aq) + H_2S(aq)$

The Law of Multiple Equilibria can be used to determine K for the above reaction.

$FeS(s) \rightleftharpoons Fe^{2+}(aq) + S^{2-}(aq);$ $K_I = K_{sp}FeS$

$H^+(aq) + S^{2-}(aq) \rightleftharpoons HS^-(aq);$ $K_{II} = 1/K_aHS^-$

$H^+(aq) + HS^-(aq) \rightleftharpoons H_2S(aq);$ $K_{III} = 1/K_aH_2S$

$FeS(s) + 2H^+(aq) \rightleftharpoons Fe^{2+}(aq) + H_2S(aq);$ $K = K_I \times K_{II} \times K_{III}$

$$K = \frac{1 \times 10^{-18}}{(1 \times 10^{-15})(1 \times 10^{-7})} = 1 \times 10^4 = \frac{[Fe^{2+}][H_2S]}{[H^+]^2}$$

$$[Fe^{2+}] = \frac{(1 \times 10^4)(1 \times 10^{-7})^2}{0.1} = 1 \times 10^{-9}$$

UNIT 21

Complex Ions: Coordination Compounds

The goal of this unit is to consider the composition, geometry, and electronic structure of complex ions. Factors that relate to the stability and rate of formation of complexes will also be considered.

INSTRUCTIONAL OBJECTIVES

Introduction
1. Define the terms complex ion, ligand, coordination number, central atom, and coordination compound.
2. Recognize that the formation of coordinate covalent bonds results from a reaction between a Lewis base and a Lewis acid.

21.1 Structures of Coordination Compounds; Charges of Complex Ions
  *1. Relate the structure of a coordination compound to its conductivity in water solution. (Prob. 21.1)
  *2. Determine the charge on a complex and on the central atom, given the formula of a coordination compound. (Probs. 21.1, 21.25b)

21.2 Composition of Complex Ions
  1. Relate the stability of a complex ion to the charge density of the metal ion and the base strength of the ligand.
  2. Realize that certain ligands can act as a chelating agent to form more than one bond with a metal ion.
  *3. Determine the coordination number of the central atom in any complex, given the formula of the species.

245

21.3 Geometry of Complex Ions
*1. Predict the geometry and draw a structural formula for a complex ion, given its structural formula. (Probs. 21.2, 21.30)
*2. Draw structural formulas to represent all the geometrical isomers of octahedral and square planar complexes, knowing the composition of the complex. (Probs. 21.2, 21.30)
3. Describe the structural difference between *cis* and *trans* isomers and methods of distinguishing between a *cis-trans* pair.

21.4 Electronic Structure of Complex Ions
1. Describe the bonding in a complex ion using valence bond theory.
2. Review the concepts of hybridization and orbital diagrams (Chapter 8 in the text).
*3. Relate the coordination numbers 2, 4, and 6 to hybridizations of sp, sp^3 or dsp^2, and d^2sp^3 orbitals. (Probs. 21.3, 21.33)
*4. Draw orbital diagrams that show the distribution of electrons around the central atom, given the formula and the geometry or having predicted the geometry. (Probs. 21.3, 21.33)
5. Relate magnetic properties of a species to the structure of the complex.
6. Summarize the essential features of the crystal-field theory.
*7. Using the crystal field approach, draw an orbital diagram for any octahedral complex, distinguishing between high spin and low spin complexes. (Prob. 21.4)
8. Relate the color of a metal ion to the extent that a ligand splits the d orbitals.

21.5 Rate of Complex Ion Formation
1. Define the terms labile and inert.
2. Distinguish between kinetic stability and thermodynamic stability.

21.6 and 21.7 Complex Ion Equilibria and Complex Ions in Analytical Chemistry
*1. Write net ionic equations for reactions involving the formation of complex ions. (Prob. 21.41)
*2. Relate the concentrations of metal ions, ligands, and complex ions, given the dissociation constant for a complex ion. (Prob. 21.5)
*3. Use the Law of Multiple Equilibria to calculate K for a reaction in which an insoluble compound is dissolved in a solution of complexing agent, given the K_{sp} and the dissociation constant for the complex ion. (Prob. 21.39)

SUMMARY

1. Structure and Bonding of Complex Ions

Complex ions are formed from a metal ion and one or more ligands. Ligands are molecules or anions that can donate an unshared electron pair to an empty orbital of the central ion to form a coordinate covalent bond. In general, ligands that are strong Lewis bases form stable complexes. The central metal ions that have a high charge density also form stable complexes. Chelating agents are ligands that can form more than one bond with a metal ion. Many biologically important compounds are chelates in which a central metal atom is bonded into a large organic molecule.

The geometry of a complex is related to the coordination number, i.e., the number of bonds formed by the central metal ion. For complexes with a coordination number of two, the geometry is linear. Those with a coordination number of six are octahedral. When the coordination number is four, the geometry may be either tetrahedral or square planar, depending on the oribtals available in the central metal ion. Geometrical isomerism can occur in octahedral complexes. Square planar complexes with the general formula MA_2B_2, where A and B are different ligands and M the metal ion, may also exhibit geometrical isomerism. Complexes with a tetrahedral geometry do not form geometrical isomers.

The valence bond approach can be used to describe and rationalize the geometries of complex ions. The relationships between coordination number, geometry, and hybridized orbitals are summarized as follows:

Coord No	Geometry	Hybridization
2	linear	sp
4	tetrahedral	sp^3
4	square planar	dsp^2
6	octahedral	d^2sp^3

The crystal field theory considers that the attractive forces between ligands and the metal ion are primarily electrostatic rather than covalent. This model describes the effect of the ligand on the energies of the orbitals of the metal ion. In an octahedral complex, ligands split the d orbitals into a low energy set of three orbitals and a high energy set of two orbitals. The magnitude of the energy difference between the two sets is influenced by the nature of the ligands. Ligands may be listed in a spectrochemical series based on their tendency to split d orbitals of a metal ion. The splitting of the d orbitals provides an explanation for properties such as color, relative stability, and behavior in a magnetic field.

Exercise 21.1

1. Complete the following table.

Formula of Complex	Metal Ion	Coord No	Ligands: Kind & No	Geometry of Complex	Hybridization of the Metal Ion
a. $[Co(NH_3)_4ClBr]^+$					
b.	Ni^{2+}		$2Br^-, 2NH_3$		dsp^2
c. $CuCl_2^-$					
d.	Zn^{2+}		$OH^-, 3H_2O$		sp^3

2. Which of the above could exhibit geometrical isomerism? Draw structures for these isomers.
3. Draw orbital diagrams to indicate the electronic structure around the central ion for the complexes in (a), (b), and (c).
4. Using crystal field theory, explain why $Cr(H_2O)_6^{3+}$ is a stable species in water solutions while $Cr(H_2O)_6^{2+}$(aq) is unstable.

Exercise 21.1 Answers

a.	Co^{3+}	6	$4NH_3, 1Br, 1Cl$	octahedral	d^2sp^3
b. $Ni(NH_3)_2Br_2$		4		square planar	
c.	Cu^+	2	$2Cl^-$	linear	sp
d. $[Zn(H_2O)_3OH]^+$		4		tetrahedral	

2. (a) and (b). $CuCl_2^-$ is linear, geometrical isomerism does not exist in complexes with tetrahedral geometry (d).

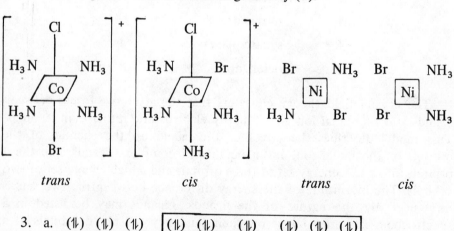

trans   cis   trans   cis

3. a. (⇅) (⇅) (⇅) [(⇅) (⇅) (⇅) (⇅) (⇅) (⇅)]
        3d   4s   4p

   b. (⇅) (⇅) (⇅) (⇅) [(⇅) (⇅) (⇅) (⇅)] ( )
        3d   4s   4p

c. (↑↓) (↑↓) (↑↓) (↑↓) (↑↓) |(↑↓)   (↑↓)| ( ) ( )
         3d   4s   4p

4. The three 3d electrons in Cr^{3+} provide a situation to maximize metal ion–ligand interaction. The Cr^{2+} with four-three d electrons, one of which is a "dz^2" or "dX^2-Y^2", would increase metal ion–ligand repulsions.

2. Complex Ions in Chemical Reactions

The kinetic stabilities of complex ions differ greatly, as indicated by the variation in rates at which they undergo substitution reactions. The thermodynamic stabilities of complex ions are related to their dissociation constants. Reactions involving complex ion equilibria are used in quantitative analysis to separate and identify metal ions. Some chelating agents, such as the sodium salt of EDTA, form extremely stable complexes with certain metal ions. These agents can be used in quantitative analysis for metal ion titrations.

Exercise 21.2
1. Write a net ionic equation for the formation of $Ag(NH_3)_2^+$ from NH_3 and the metal ion in solution.
2. Calculate the ratio of $[Ag^+]/[Ag(NH_3)_2^+]$ in a solution when the equilibrium concentration of NH_3 is 1.0×10^{-2} M.

(Kd $Ag(NH_3)_2^+ = 4.0 \times 10^{-8}$)

3. Write a net ionic equation and calculate K for the reaction between AgI and NH_3(aq).

(K_{sp} AgI = 1.0×10^{-16})

4. Would you expect AgI to be appreciably soluble in NH_3 solution? If not, refer to a table of dissociation constants (Table 21.6) and select a more suitable complexing agent.
5. A sample of iron ore weighing 0.756g is brought into solution and titrated with EDTA. It requires 22.0 ml of 0.110 M EDTA. Calculate the percentage of Fe^{3+} in the sample.

Exercise 21.2
1. $Ag^+(aq) + 2NH_3(aq) \rightleftharpoons Ag(NH_3)_2^+ (aq)$

2. $Ag(NH_3)_2^+(aq) \rightleftharpoons Ag^+(aq) + 2NH_3(aq)$

$$Kd = \frac{[Ag^+][NH_3]^2}{[Ag(NH_3)_2^+]} ; \frac{[Ag^+]}{[Ag(NH_3)_2^+]} = \frac{Kd}{[NH_3]^2} = \frac{4.0 \times 10^{-8}}{(1.0 \times 10^{-2})^2} =$$

4.0×10^{-4}

3. $AgI(s) \rightarrow Ag^+(aq) + I^-(aq)$ $\qquad K_1 = K_{sp}$

   $Ag^+(aq) + 2NH_3(aq) \rightarrow Ag(NH_3)_2^+(aq)$ $\qquad K_2 = 1/K_d$

   $AgI(s) + 2NH_3(aq) \rightarrow Ag(NH_3)_2^+(aq) + I^-(aq)$ $\qquad K$

   $$K = \frac{K_{sp}AgI}{K_d Ag(NH_3)_2^+} = \frac{1.0 \times 10^{-16}}{4.0 \times 10^{-8}} = 2.5 \times 10^{-9}$$

4. No, the value of K for the reaction is very small. A better complexing agent would be CN^-. The K_d for $Ag(CN)_2^-$ is much smaller. The K for the reaction using CN^- is 1×10^5.

5. Number moles Fe^{3+} = number moles EDTA

   $$0.110 \, \frac{\text{mole EDTA}}{\text{liter}} \times 0.0220 \text{ liter} \times \frac{1 \text{ mole Fe}^{3+}}{1 \text{ mole EDTA}} \times \frac{55.8g}{\text{mole Fe}^{3+}} \times \frac{1}{0.756g} \times 100 = 17.9\%$$

SUGGESTED ASSIGNMENT XXI

Text: Chapter 21

Problems: Numbers 21.1–21.5, 21.25, 21.30, 21.33, 21.39, 21.41

Solutions to Assigned Problems Set XXI

21.1

The fact that the complex ion has an octahedral geometry indicates that the coordination number is six. The complex ion is $Cr(NH_3)_4Br_2^+$. Two of the three Br^- along with the four NH_3 molecules are bonded directly to Cr^{3+}. Since two Br^- and four neutral NH_3 are bonded to the Cr^{3+}, the net charge on the complex is $+3+4(0)+2(-1)=+1$. The formula of the compound is $[Cr(NH_3)_4Br_2]Br$. This would show a conductivity similar to that of NaCl.

21.2

$Cu(H_2O)_2^+$: The coordination number is two and the geometry is linear. $[H_2O-Cu-OH_2]^+$

$Zn(H_2O)_2Br_2$: The coordination number is four and a tetrahedral geometry is indicated. Note that no isomers are possible with a tetrahedral geometry.

$$\begin{array}{c} H_2O \\ | \\ Zn \\ \diagup \ | \ \diagdown \\ H_2O \quad \ \ \ Br \\ Br \end{array}$$

$Cu(H_2O)_2Br_2$: This is a complex that contains Cu^{2+} as the central metal ion. Its geometry is square planar. Two isomers are possible.

$$\begin{array}{cc} Br \quad \ \ Br & Br \quad \ \ OH_2 \\ \diagdown \diagup & \diagdown \diagup \\ Cu & Cu \\ \diagup \diagdown & \diagup \diagdown \\ H_2O \quad \ \ OH_2 & H_2O \quad \ \ Br \\ \textit{cis} & \textit{trans} \end{array}$$

$Fe(H_2O)_5Br^{2+}$: The central metal ion is Fe^{3+}. The geometry is octahedral.

$$\left[\begin{array}{c} H_2O \\ H_2O - Fe - H_2O \\ H_2O \quad H_2O \\ Br \end{array}\right]^{2+}$$

21.3

Since the coordination number is six, we can predict that the bonding electrons contributed by the ligands will enter d^2sp^3 hybrid orbitals. The central metal ion must have a charge of +3 based on the reasoning that the Br^- has a charge of -1, H_2O is 0, and the complex has a charge of +2.

Locating the bonding electrons first, we have:

```
      3d              4s         4p
( )( )( )(↑↓)(↑↓)    (↑↓)    (↑↓)(↑↓)(↑↓)
```

Now, we insert the electrons associated with Fe^{3+}. Since Fe has atomic number 26, the Fe^{3+} now would leave $26 - 3 = 23$ electrons. Of these, 18 are in the Ar core; the remaining 5 enter the available 3d orbitals in accordance with Hund's rule. The final diagram is:

```
      3d              4s         4p
(↑↓)(↑↓)(↑) (↑↓)(↑↓)  (↑↓)   (↑↓)(↑↓)(↑↓)
```

21.4

In Problem 21.3, we determined that there were five d electrons in Fe^{3+}. With strongly interacting ligands, the five d electrons would be repelled into the lower three energy levels forming a low spin complex. They would be spread out over the five d orbitals according to Hund's rule to form a high spin complex with weak interacting ligands.

21.5

Assume a simple one step dissociation:

$Zn(NH_3)_4^{2+}(aq) \rightleftharpoons Zn^{2+}(aq) + 4NH_3(aq)$

$$Kd = \frac{[Zn^{2+}][NH_3]^4}{[Zn(NH_3)_4^{2+}]} ; \frac{[Zn^{2+}]}{[Zn(NH_3)_4^{2+}]} = \frac{1 \times 10^{-9}}{(6)^4} = 8 \times 10^{-13}$$

21.25

a. The ligands are Cl^-, and H_2O which is neutral.

b. Since there is only one Br^-, the complex must have a positive one charge. The H_2O is neutral and there is one chloride ion; therefore, the central metal ion must be Fe^{2+}.

c. One mole of Ag^+ would precipitate one mole of Br^-. The Cl^- is bonded to the Fe^{2+} and would not readily precipitate at this temperature.

21.30

a. The coordination number is six, so we predict an octahedral geometry for $Cr(NH_3)_5 Br^{2+}$

$$\left[\begin{array}{c} NH_3 \\ NH_3 \diagup \stackrel{|}{Cr} \diagdown NH_3 \\ NH_3 \stackrel{|}{} NH_3 \\ Br \end{array} \right]^{2+}$$

b. $Cu(H_2O)_3 Cl^+$

$$\left[\begin{array}{c} H_2O \diagdown \diagup OH_2 \\ Cu \\ H_2O \diagup \diagdown Cl \end{array} \right]^+$$

c. *cis*-Co(NH$_3$)$_4$BrCl$^+$

$$\left[\begin{array}{c} \text{NH}_3 \\ \text{NH}_3-\text{Co}-\text{Br} \\ \text{NH}_3 \quad \text{Cl} \\ \text{NH}_3 \end{array}\right]^+$$

d. *trans*-Ni(H$_2$O)$_2$Cl$_2$ This must be square planar since the trans isomer is indicated. Geometrical isomers do not exist when a complex has a tetrahedral geometry.

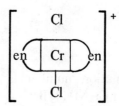

e. *trans*-Cr(en)$_2$Cl$_2$$^+$

$$\left[\text{en}-\text{Cr}\begin{array}{c}\text{Cl}\\ \\ \text{Cl}\end{array}\text{en}\right]^+$$

f. Pt(en)$_2$$^{2+}$

$$\left[\text{en}-\text{Pt}-\text{en}\right]^{2+}$$

21.33

a. The coordination number is two. The bonding electrons from NH$_3$ should enter sp hybrid orbitals. In Ag$^+$, these are the 5s and 5p orbitals.

```
        4d                              3s          5p
  ⇅  ⇅  ⇅  ⇅  ⇅                         ⇅     ⇅   __  __
```

b. Referring to Table 21.3, we see that Cu^{2+} with a coordination

UNIT 21—COMPLEX IONS: COORDINATION COMPOUNDS • 255

number of four forms square planar complexes. Square planar complexes have a dsp² hybridization.

```
         3d                    4p
   ⇅  ⇅  ⇅  ⇅  ⇅      ⇅      ⇅  ⇅  ↿
```

Note that the unpaired electron is moved to an unhybridized 4p orbital.

c. By reasoning similar to that in Problems 21.1 and 21.3, we conclude that the central ion is Co^{3+}. The coordination number is six and the hybridization is d^2sp^3.

```
         3d              4s              4p
   ⇅  ⇅  ⇅  ⇅  ⇅        ⇅         ⇅  ⇅  ⇅
```

d. It is obvious that this is Co^{2+}, with a coordination number of six. The hybridization would be d^2sp^3 with a lone electron in an unhybridized 4d orbital.

```
         3d              4s        4p            4d
   ⇅  ⇅  ⇅  ⇅  ⇅        ⇅     ⇅  ⇅  ⇅     ↿  _  _  _  _
```

e. The hybridization would be d^2sp^3; there are four 3d electrons in Mn^{2+}.

```
         3d              4s        4p
   ⇅  ↿  ↿  ⇅  ⇅         ⇅      ⇅  ⇅  ⇅
```

21.39

From Table 21.6, $K_d = 2 \times 10^{-13}$. The K_{sp} for CuS from Table 18.2 is found to be 1×10^{-36}. Using the Rule of Multiple Equilibria:

$CuS(s) \rightleftharpoons Cu^{2+}(aq) + S^{2-}(aq);$ $\qquad K_1 = K_{sp}$ of CuS

$Cu^{2+}(aq) + 4NH_3(aq) \rightleftharpoons Cu(NH_3)_4^{2+}(aq);$ $\qquad K_2 = \dfrac{1}{K_d}$

$$CuS(s) + 4NH_3(aq) \rightleftharpoons Cu(NH_3)_4^{2+}(aq) + S^{2-}(aq); K = K_1 \times K_2 = \frac{K_{sp}}{K_d};$$

$$K = \frac{1 \times 10^{-36}}{2 \times 10^{-13}} = 5 \times 10^{-24}$$

Since K is very small, aqueous NH_3 would not be an effective solvent for CuS.

21.41

a. $Cu(NH_3)_4^{2+}(aq) + 4H^+(aq) + 4\ Cl^-(aq) \rightleftharpoons CuCl_4^{2-}(aq) + 4NH_4(aq)$

b. $Fe^{3+}(aq) + 3OH^-(aq) \rightarrow Fe(OH)_3(s)$

$Zn^{2+}(aq) + 4OH^-(aq) \rightarrow Zn(OH)_4^{2-}(aq)$

The $Fe(OH)_3$ is not amphoteric; that is, it does not react with excess OH^- to form a soluble complex ion.

c. $Fe(H_2O)_6^{3+}(aq) + SCN^-(aq) \rightarrow Fe(H_2O)_5SCN^{2+}(aq) + H_2O$

UNIT 22

Oxidation and Reduction: Electrochemical Cells

The goal of this unit is to consider some fundamental concepts of oxidation-reduction and apply these concepts to electrolytic and voltaic cells.

INSTRUCTIONAL OBJECTIVES

22.1 Oxidation Number
1. Distinguish between oxidation and reduction in terms of loss and gain of electrons.
2. Learn the rules for assigning oxidation numbers.
*3. Determine the oxidation number of each atom, given the formula of an ion or molecule. (Probs. 22.1, 22.25)
4. Distinguish between oxidation and reduction in terms of change in oxidation number. (Prob. 22.29a, b)
5. Distinguish between an oxidizing agent and a reducing agent. (Prob. 22.29c, d)

22.2 Balancing Oxidation-Reduction Equations
*1. Balance a redox equation using the half-equation method, given the formulas of the reactants and products. (Probs. 22.2, 22.33)
*2. Relate quantities of products and reactants, given or having derived the equation for a redox reaction. (Prob. 22.35)

22.3 Electrolytic Cells
1. Define and illustrate, with reference to an electrolytic cell, oxidation, reduction, anode, and cathode.

257

*2. Describe, using labeled diagrams, the physical construction and operation of an electrolytic cell. (Prob. 22.4a)
*3. Given any two of the following for an electrolytic cell, calculate the third: (a) current in amperes, (b) time of current flow through the cell, and (c) quantity of substances produced or consumed. (Probs. 22.3, 22.39)
4. Recognize the differences between electrolysis in aqueous solution and electrolysis of molten salts.
*5. Relate the GEW of a species to its gram formula weight, given the quantities involved in a redox reaction. (Prob. 22.42)
6. Realize that electrolytic reactions are nonspontaneous, $\Delta G > 0$.

22.4 Voltaic Cells
1. Realize that voltaic cell reactions are spontaneous, $\Delta G < 0$.
2. Define and illustrate, with reference to a voltaic cell, oxidation, reduction, anode, cathode, and salt bridge.
*3. Describe, using a labeled diagram, the physical construction and operation of a voltaic cell. (Prob. 22.4b)
4. Describe the operation of a fuel cell.

SUMMARY

1. Oxidation Numbers and Redox Equations

Redox reactions involve the transfer of electrons from a reducing agent to an oxidizing agent. As a result of the electron transfer, the reducing agent increases in oxidation number and the oxidizing agent exhibits a decrease in oxidation number. Oxidation may be defined as an increase in oxidation number; reduction is a decrease in oxidation number. Oxidation numbers are assigned according to a set of arbitrary rules. Exercise 22.1 illustrates the application of these rules.

Balancing equations by the half-equation method involves separating the ionic equation for the oxidation reaction from the ionic equation for the reduction reaction. Each equation is balanced for mass and charge. The balanced half-equations are then adjusted so that the number of electrons gained is equal to the number of electrons lost. When the two half-equations are added, no electrons appear in the resulting net equation.

Exercise 22.1
1. Give the oxidation number of each atom in:

$K_3 VO_4$; $Ca(ClO_3)_2$; $Na_2 S_2 O_3$; $Cr_2 O_7^{2-}$

2. Consider the following unbalanced equation for a reaction in acid solution.

$$H_2SO_3(aq) + MnO_4^-(aq) \rightarrow SO_4^{2-}(aq) + Mn^{2+}(aq)$$

   a. Write a balanced half-equation for the oxidation reaction.
   b. Write a balanced half-equation for the reduction reaction.
   c. Write a balanced net ionic equation for the overall reaction.
   d. Give the formula of the reducing agent.
3. Balance the equation for the following reaction in basic solution.

$$MnO_4^-(aq) + CN^-(aq) \rightarrow MnO_2(s) + CNO^-(aq)$$

4. A 1.000g sample of an alloy containing tin is treated with nitric acid to produce 0.573g of $SnO_2(s)$. The NO_3^- is reduced to $NO(g)$.
   a. Write a balanced equation for the reaction.
   b. Determine the percentage of tin in the sample.
   c. Calculate the number of moles of H^+ consumed in the reaction.

Exercise 22.1 Answers

1. This exercise calls for the application of the rules for assigning oxidation numbers.

   For K_3VO_4: $K = +1; O = -2; V = +5$

   $Ca(ClO_3)_2$: $Ca = +2; O = -2; Cl = +5$

   $Na_2S_2O_3$: $Na = +1; O = -2; S = +2$

   $Cr_2O_7^{2-}$: $O = -2; Cr = +6$

2. a. oxidation half-reaction:

   $$H_2SO_3(aq) + H_2O \rightarrow SO_4^{2-}(aq) + 4H^+(aq) + 2e^-$$

   b. reduction half-reaction:

   $$MnO_4^-(aq) + 8H^+(aq) + 5e^- \rightarrow Mn^{2+}(aq) + 4H_2O$$

   c. net ionic equation.

   $$2MnO_4^-(aq) + 5H_2SO_3(aq) \rightarrow 2Mn^{2+}(aq) + 5SO_4^{2-}(aq) + 4H^+(aq) + 3H_2O$$

   d. H_2SO_3 is the reducing agent

3. $2(MnO_4^-(aq) + 4H^+(aq) + 3e^- \to MnO_2(s) + 2H_2O)$

   $\underline{3(CN^-(aq) + H_2O \to CNO^-(aq) + 2H^+(aq) + 2e^-)}$

   $2MnO_4^-(aq) + 2H^+(aq) + 3CN^-(aq) \to 2MnO_2(s) + H_2O + 3CNO^-$

   $\underline{+2OH^- \to +2OH^-}$

   $2MnO_4^-(aq) + H_2O + 3CN^-(aq) \to 2\ MnO_2(s) + 3CNO^-(aq) + 2OH^-(aq)$

4. a. $3Sn(s) + 4NO_3^-(aq) + 4H^+(aq) \to 3SnO_2(s) + 4NO(g) + 2H_2O$

   b. $0.573g\ SnO_2 \times \dfrac{118.7g\ Sn}{150.7g\ SnO_2} \times \dfrac{1}{1.000g} \times 100 = 45.1\%$

   c. $0.573g\ SnO_2 \times \dfrac{1\ mole\ SnO_2}{150.7g\ SnO_2} \times \dfrac{4\ moles\ H^+}{3\ moles\ SnO_2} = 5.07 \times 10^{-3}\ mole\ H^+$

2. Electrochemical Cells

For an electrolytic cell, electrical energy can be used to bring about nonspontaneous redox reactions. The electron transfer process that occurs in an electrolytic cell depends on the material between the anode and the cathode. The simplest processes are those that occur when the cell contains a molten salt, such as NaCl. The process is not as straightforward when the electrolytic cell contains an aqueous solution because water molecules, H^+, OH^-, and ions of the solute are present.

Faraday's Law relates the amount of electricity passed through a cell to the amounts of substances produced or consumed. The amount of charge associated with one mole of electrons is 96,500 coulombs. This amount of electrical charge is also called a faraday. The value of writing half-equations for redox reactions should be apparent since the number of moles of electrons in the ionic half-equation is the number of faradays involved in the reaction. The weight of a substance oxidized or reduced by one faraday (one mole of electrons) is the gram equivalent weight.

In electrolytic cells, electrical energy is used to bring about oxidation at the anode and reduction at the cathode. A voltaic cell has reactions occuring spontaneously at the electrodes to produce electrical energy. In principle, any spontaneous redox reaction can be used as a basis for a voltaic cell. The oxidation reaction is separated from the reduction

UNIT 22–OXIDATION AND REDUCTION: ELECTROCHEMICAL CELLS · 261

reaction in such a way that the electrons donated by the reducing agent flow through an external conductor to reach the oxidizing agent. The cell compartments must be connected, usually by a salt bridge or porous plate, so that ions can flow from one half-cell to the other. As a result of the migration of these ions, each compartment remains electrically neutral.

Another type of voltaic cell is the fuel cell. This converts the chemical energy of conventional fuels, such as hydrogen and methane, directly into electrical energy at extremely high efficiencies.

Exercise 22.2
1. Sodium metal is produced in the Downs process by the electrolysis of molten NaCl. Determine the number of hours required to produce 2299g of sodium metal if the current is 10.0 amps. Assume the process is 100% efficient.
2. When a solution of a certain lanthanum salt is electrolyzed, it is found that a current of 1.93 amperes forms 6.66g of La in two hours. Calculate the GEW of lanthanum and the oxidation number of lanthanum in the solution species.

Exercise 22.2 Answers
1. The reaction is $Na^+(l) + 1e^- \rightarrow Na(l)$

$$2299g \times \frac{1 \text{ mole}}{22.99g} \times \frac{1 \text{ F}}{1 \text{ mole}} \times \frac{96,500 \text{ coulombs}}{1 \text{ F}} = 9.65 \times 10^6 \text{ coul}$$

coul = amp-sec; 9.65×10^6 amp sec $\times \dfrac{1}{10 \text{ amp}} = 9.65 \times 10^5$ sec

9.65×10^5 sec $\times \dfrac{1 \text{ min}}{60 \text{ sec}} \times \dfrac{1 \text{ hr}}{60 \text{ min.}} = 2.68 \times 10^2$ hr

2. 1.93 amp $\times$ 7200 sec $\times \dfrac{1 \text{ F}}{96,500 \text{ coul}} = 0.144$ F = number of GEW

La

$6.66g \times \dfrac{1}{0.144 \text{ GEW}} = 46.3$ g/GEW

$\dfrac{138.9g}{\text{GAW}} \times \dfrac{1 \text{ GEW}}{46.3g} = 3$ GEW/GAW. The GEW is one-third the gram atomic weight of lanthanum. The La must have an oxidation number of +3.

SUGGESTED ASSIGNMENT XXII

Text: Chapter 22

Problems: Numbers 22.1–22.4, 22.25, 22.29, 22.33, 22.35, 22.39, 22.42

Solutions to Assigned Problems Set XXII

22.1

a. Using the rules in Section 22.1 for assigning oxidation numbers, we know that the oxidation number of sodium must be +1. The charge on the phosphate must be −3. Taking the oxidation of oxygen to be −2, phosphorus must be:

oxid number P + 4 (−2) = −3; oxid number P = +5

b. We recognize that the sum of the oxidation numbers of all atoms is zero in a neutral species. Assigning each H an oxidation number of +1, each carbon must be:

4 (oxid number C) + 10(+1) = 0; oxid number C = −2½

c. Taking the oxidation number of oxygen to be −2, and recognizing that the sum of the oxidation numbers is −1, N must be:

oxid number N + 3(−2) = −1; oxid number N = +5

22.2

Using the half-equation method, we separate the equation into an oxidation half-equation and a reduction half-equation, then balance each half-equation with respect to mass. For reactions in acid solution, H_2O and H^+ are added to balance the mass. Electrons are added to balance the charge. Finally, each half-equation is multiplied by a factor that will result in a conservation of the number of electrons in each half-equation. When the equations are added, no electrons will appear on either side of the final equation.

oxidation: $Fe(s) \rightarrow Fe^{3+}(aq) + 3e^-$

reduction: $2e^- + 4H^+(aq) + SO_4^{2-}(aq) \rightarrow SO_2(g) + 2H_2O$

To conserve electrons, multiply the oxidation half-equation by 2; the reduction half-equation by 3. Add the two half-equations and simplify.

$$2 (Fe(s) \rightarrow Fe^{3+}(aq) + 3e^-)$$

$$\underline{3 (2e^- + 4H^+(aq) + SO_4^{2-}(aq) \rightarrow SO_2(g) + 2H_2O)}$$

$$2Fe(s) + 12H^+(aq) + 3SO_4^{2-}(aq) \rightarrow 2Fe^{3+}(aq) + 3SO_2(g) + 6H_2O$$

Note that the oxidation half-equation always contains electrons on the right side of the equation; while the reduction half-equation contains electrons on the left side.

22.3

The equation for the process is

$$2H_2O + 2Cl^-(aq) \rightarrow Cl_2(g) + H_2(g) + 2OH^-(aq)$$

Using Equation 22.17, we can calculate the number of coulombs.

$$2 \text{ hours} \times \frac{60 \text{ min}}{\text{hour}} \times \frac{60 \text{ sec}}{\text{min}} \times 10.0 \text{ amperes} = 72,000 \text{ coulombs}$$

We note that the formation of each mole of Cl_2 involves two moles of electrons or two faradays. By Equation 22.16, 1 faraday = 96,500 coulombs. Using conversion factors,

$$72,000 \text{ coulombs} \times \frac{1 \text{ faraday}}{96,500 \text{ coulombs}} \times \frac{1 \text{ mole } Cl_2}{2 \text{ faradays}} \times \frac{70.9g \; Cl_2}{\text{mole } Cl_2}$$

$$= 26.4g \; Cl_2$$

22.4

a. The diagram would be similar to the cell shown in Figure 22.2. The reduction reaction occuring at the cathode is

$$Ca^{2+}(l) + 2e^- \rightarrow Ca(l)$$

The oxidation reaction occuring at the anode is

$$2Cl^- \rightarrow Cl_2(g) + 2e^-$$

The electrons flow from the negative terminal of the direct current source (battery) to the cathode and from a graphite electrode, which serves as the anode, back to the positive terminal of the battery. The liquid Ca^{2+} ions migrate to the cathode, where they are reduced to liquid calcium metal. The Cl^- ions in the molten $CaCl_2$ migrate to the anode, where they release electrons to form gaseous Cl_2.

b. This diagram would be similar to the voltaic cell shown in Figure 22.6. A salt bridge containing a solution of KNO_3 connects the anode compartment with the cathode compartment. A strip of Pb immersed in a $Pb(NO_3)_2$ solution serves as the anode in this half-cell. Electrons flow from the lead electrode through the external circuit when the reaction $Pb(s) \rightarrow Pb^{2+}(aq) + 2e^-$ occurs in the anode compartment. In the cathode compartment, silver ions from an $AgNO_3$ solution move to the silver metal cathode where the reaction $Ag^+(aq) + e^- \rightarrow Ag(s)$ occurs. Note that two silver atoms form for every Pb atom that is oxidized. As current is drawn from the cell, negative ions move to the anode, positive ions to the cathode. Electrical neutrality is maintained throughout the system.

22.25

The reasoning here is similar to that in Problem 22.1.

a. Assigning each oxygen a -2 oxidation number,

$3(-2) + 2(\text{oxid number As}) = 0$; $As = +3$

b. Calcium, a Group 2A element, has a $+2$ oxidation number in a compound. The oxalate ion must have a charge of -2. Assigning oxygen a -2 oxidation number,

$2(\text{oxid number C}) + 4(-2) = -2$; oxid number $C = +3$

c. Fluorine has an oxidation number of -1 in all of its compounds.

$5(\text{oxid number P}) + 5(-1) = 0$; oxid number $P = +5$

d. Assigning oxygen an oxidation number of -2,

$(\text{oxid number N}) + 2(-2) = -1$; oxid number $N = +3$

e. Assigning oxygen an oxidation number of -2,

$2(\text{oxid number Cr}) + 7(-2) = -2$; oxid number Cr = +6

f. Oxygen does not exhibit its usual oxidation number here. The oxidation number is $-\frac{1}{2}$.

22.29

a. Chlorine is oxidized from an oxidation state -1 to zero.

b. Manganese goes from +7 to Mn^{2+}. It is reduced.

c. The MnO_4^- is the oxidizing agent, it brings about the oxidation of the Cl^-.

d. The Cl^- is the reducing agent, it brings about the reduction of MnO_4^- to Mn^{2+}.

22.33

Use the same approach as in Problem 22.2.

a. Note that Cl_2 is being both oxidized and reduced.

oxidation: $8H_2O + Cl_2(g) \rightarrow 2ClO_4^-(aq) + 16H^+ + 14e^-$

reduction: $\quad 2e^- + Cl_2(g) \rightarrow 2Cl^-(aq)$

Multiply the reduction half-equation by 7 and add the two equations.

$8H_2O + Cl_2(g) + 14e^- + 7Cl_2(g) \rightarrow 2ClO_4^-(aq) + 16H^+(aq) + 14e^- + 14Cl^-(aq)$

Simplifying: $4H_2O + 4Cl_2(g) \rightarrow ClO_4^-(aq) + 8H^+(aq) + 7Cl^-(aq)$

b. Even though this reaction occurs in basic solution, proceed as though it occurs in acid solution to obtain the same equation as in (a). Then add $8OH^-$ to each side of the equation, forming eight molecules of H_2O on the right side of the equation.

$4H_2O + 4Cl_2(g) \rightarrow ClO_4^-(aq) + 8H^+(aq) + 7Cl^-(aq)$

$\underline{\qquad\qquad + 8OH^- \quad\rightarrow\qquad\qquad + 8OH^-\qquad\qquad}$

$4H_2O + 4Cl_2(g) + 8OH^- \rightarrow ClO_4^-(aq) + 8H_2O + 7Cl^-(aq)$

Subtract four moles of H_2O from each side to give:

$4Cl_2(g) + 8OH^-(aq) \rightarrow ClO_4^-(aq) + 4H_2O + 7Cl^-(g)$

c. oxidation: $3H_2O + AsH_3(g) \rightarrow H_3AsO_3(aq) + 6H^+(aq) + 6e^-$

   reduction: $3e^- + AuCl_4^-(aq) \rightarrow Au(s) + 4Cl^-(aq)$

Multiply the reduction half-equation by 2, add it to the oxidation half-equation, and simplify.

$3H_2O + AsH_3(g) + 2AuCl_4^-(aq) \rightarrow H_3AsO_3(aq) + 6H^+(aq) + 2Au(s) + 8Cl^-(aq)$

d. Determine the overall equation as though the reaction occurs in acid solution. Then add the same number of OH^- to each side as there are H^+ and simplify.

   oxidation: $2H_2O + Am^{3+}(aq) \rightarrow AmO_2^+(aq) + 4H^+(aq) + 2e^-$

   reduction: $2e^- + S_2O_8^{2-}(aq) \rightarrow 2SO_4^{2-}(aq)$

$2H_2O + Am^{3+}(aq) + S_2O_8^{2-}(aq) \rightarrow AmO_2^+(aq) + 4H^+(aq) + 2SO_4^{2-}(aq)$

$+ 4OH^- \quad\quad\quad \rightarrow \quad\quad\quad + 4OH^-$

$Am^{3+}(aq) + S_2O_8^{2-}(aq) + 4OH^-(aq) \rightarrow AmO_2^+(aq) + 2H_2O + 2SO_4^{2-}(aq)$

22.35

a. oxidation: $Zn(s) \rightarrow Zn^{2+}(aq) + 2e^-$

   reduction: $8e^- + 11H^+(aq) + AsO_4^{3-}(aq) \rightarrow AsH_3(g) + 4H_2O$

Multiply the oxidation half-equation by 4, add to the reduction half-equation, and simplify.

$4Zn(s) + 11H^+(aq) + AsO_4^{3-}(aq) \rightarrow 4Zn^{2+}(aq) + AsH_3(g) + 4H_2O$

b. Using conversion factors,

$1.00\text{g AsO}_4^{3-} \times \dfrac{1 \text{ mole AsO}_4^{3-}}{139\text{g AsO}_4^{3-}} \times \dfrac{11 \text{ moles H}^+}{1 \text{ mole AsO}_4^{3-}} = 0.0791$ mole H^+

c. This problem is similar to (b), in that we calculate the number of moles of AsH_3 produced. We then use the Ideal Gas Law to calculate the volume of AsH_3 at these conditions.

$$1.00 \times 10^{-3} \text{g } AsO_4^{3-} \times \frac{1 \text{ mole } AsO_4^{3-}}{139 \text{ g } AsO_4^{3-}} \times \frac{1 \text{ mole } AsH_3}{1 \text{ mole } AsO_4^{3-}} = 7.19 \times 10^{-6} \text{ mole } AsH_3$$

$$V = \frac{nRT}{P} = \frac{(7.19 \times 10^{-6} \text{ mole})\left(\frac{0.0821 \text{ liter atm}}{°K \text{ mole}}\right)(298°K)}{1 \text{ atm}}$$

$$= 1.76 \times 10^{-4} \text{ liter} = 1.76 \times 10^{-1} \text{ cm}^3$$

22.39

a. The reduction reaction is $Al^{3+} + 3e^- \to 3Al(s)$

Using conversion factors,

$$\frac{1.00 \times 10^3 \text{g Al}}{\text{day}} \times \frac{1 \text{ mole Al}}{27.0 \text{g Al}} \times \frac{3 \text{ moles } e^-}{1 \text{ mole Al}} \times \frac{6.02 \times 10^{23} e^-}{\text{mole } e^-} = 6.69 \times 10^{25} e^-/\text{day}$$

b. We can first calculate the number of coulombs, then the number of seconds in a day.

$$1.00 \times 10^3 \text{g Al} \times \frac{1 \text{ mole Al}}{27.0 \text{ g Al}} \times \frac{3 \text{ moles } e^-}{1 \text{ mole Al}} \times \frac{96,500 \text{ coulombs}}{\text{mole } e^-} = 1.07 \times 10^7 \text{ coulombs}$$

$$24 \text{ hr} \times \frac{3600 \text{ sec}}{\text{hr}} = 8.64 \times 10^4 \text{ sec}$$

From Equation 22.17, number amp = number coulombs/number seconds =

$1.07 \times 10^7 \text{ coul}/8.64 \times 10^4 \text{ sec} = 1.24 \times 10^2 \text{ amp}$

c. The equation for the reaction is

$2Al^{3+} + 3O^{2-} \to 2Al(s) + 1.5O_2(g)$

Using conversion factors,

$$1.00 \times 10^3 \text{g Al} \times \frac{1 \text{ mole Al}}{27.0 \text{g Al}} \times \frac{1.5 \text{ moles O}_2}{2 \text{ moles Al}} \times \frac{32.0 \text{g}}{\text{mole O}_2} = 8.89 \times 10^2 \text{g}$$

22.42

a. We can convert the 4800 coulombs to the number of faradays and then calculate the weight of platinum deposited per faraday.

$$4800 \text{ coulombs} \times \frac{1 \text{ faraday}}{96,500 \text{ coul}} = 4.97 \times 10^{-2} \text{ faraday}$$

$$2.44 \text{g} \times \frac{1}{4.97 \times 10^{-2} \text{ faraday}} = 49.1 \text{g/faraday}$$

The GEW = number of grams per faraday = 49.1

b. The GAW of Pt is 195. The number of GEW per one GAW is

$$195 \frac{\text{g}}{\text{GAW}} \times \frac{1 \text{ GEW}}{49.1 \text{g}} = 3.97 \approx 4 \frac{\text{GEW}}{\text{GAW}}$$

This implies that the Pt species in solution has an oxidation number of +4. The reduction half-equation is

$$Pt^{4+}(aq) + 4e^- \rightarrow Pt(s)$$

UNIT 23

Oxidation-Reduction Reactions: Spontaneity and Extent

The goal of this unit is to apply the principles of voltaic cells to determine the spontaneity and extent of redox reactions and to consider some applications of these reactions.

INSTRUCTIONAL OBJECTIVES

23.1 Standard Potentials
1. Define: standard voltage, $E°$, standard reduction potential, and standard cell conditions.
2. Realize that the voltage of a cell at a given temperature is dependent upon the nature of the cell reaction and the concentrations of the species involved in the reaction.
3. Recognize that the standard cell voltage is the sum of the standard reduction potential and standard oxidation potential.
4. Know that the standard reduction potential for hydrogen is used as a reference and arbitrarily assigned a value of zero.
5. Recognize that the standard potentials for the forward (reduction) half-reaction and reverse (oxidation) half-reaction are equal in magnitude but opposite in sign.
6. Use standard electrode potentials to:
   *a. compare relative strengths of different reducing agents; different oxidizing agents. (Probs. 23.2, 23.28)
   *b. calculate a cell voltage at standard conditions ($E°$). (Probs. 23.1, 23.24)

23.2 Spontaneity and Extent of Redox Reactions
*1. Use standard electrode potentials to predict whether a particular redox reaction will occur spontaneously at standard conditions.
*2. Relate the standard free energy change to the standard voltage: $\Delta G° = -nFE°$. (Probs. 23.3, 23.31)
*3. Relate the standard cell potential to the equilibrium constant for a redox reaction: $\log_{10} K = nE°/0.0591$. (Probs. 23.3, 23.31)
*4. Calculate K, given or having determined $E°$ or $\Delta G°$.
*5. Use K values to determine the position of a redox equilibrium.

23.3 Effect of Concentration on Voltage
1. Use the Nernst equation to calculate:
   *a. voltage of a cell, given or having calculated $E°$ and the concentrations of all species. (Probs. 23.4a, 23.35)
   *b. potential of a half-cell, given or having calculated $E°$ and the concentrations of all species.
   *c. the concentration of one species, given or having calculated E, $E°$, and all other species. Use these concentrations to calculate equilibrium constants (e.g., K_{sp}, K_a, K_w). (Probs. 23.4b, 23.36)
2. Explain the essential features of a pH meter.

23.4, 23.5 Strong Oxidizing Agents; Oxygen, the Corrosion of Iron
1. Discuss the characteristics of some common oxidizing agents, include the effect of pH on the equilibria involved.
2. Discuss the electrochemical mechanism of the corrosion of iron, and methods that may be used to prevent corrosion.

23.6 Redox Reactions in Analytical Chemistry
1. Illustrate, by use of net ionic equations, how redox reactions are used in qualitative analysis. (Prob. 23.39)
*2. Calculate the amount or concentration of a reactant, given titration data for a redox reaction. (Probs. 23.5, 23.41)

SUMMARY

1. Cell Potentials as a Measure of the Spontaneity and Extent of Redox Reactions

The cell potential is a measure of the driving force of the redox reaction in a voltaic cell. Standard cell voltages are obtained when the temperature is 25°C and all species in solution are at one molar concentration and all gases at one atmosphere pressure. The standard hydrogen electrode is assigned a reduction potential of zero volts. Other standard reduction potentials (SRP) are compared to that of the hydrogen cell. A positive value for a standard reduction potential indicates that a reduction half-reaction has a greater tendency to occur than the reaction

$$2H^+(aq, 1\ M) + 2e^- \rightarrow H_2(g, 1\ atm)$$

A negative standard reduction potential means that a reduction half-reaction has less of a tendency to occur than the above reaction. Standard oxidation potentials (SOP) are obtained for the reverse (oxidation) half-reaction by changing the sign of the reduction potential.

The overall redox reaction is obtained by adding the reduction and oxidation half-reactions. A standard cell potential is the algebraic sum of the SOP for the species being oxidized and the SRP for the species being reduced. A positive value of a standard cell potential indicates that the reaction is spontaneous. Remember that multiplying an equation of a half-cell reaction by some factor does not change the voltage of that half-reaction.

The sign of $E°$ can be used to predict the spontaneity of a reaction. If $E°$ is positive, the reaction is spontaneous and $\Delta G°$ must be negative. The expression $\Delta G° = -nFE°$ relates the standard voltage to the change in free energy. The extent of a redox reaction is related to the value of the equilibrium constant K. The equation $\log_{10} K = \dfrac{nE°}{0.0591}$ may be used to calculate K from the standard cell voltage.

The Nernst equation can be used to calculate the voltage, E, of a cell when the species in solution in a redox reaction are present at concentrations other than one molar, or when the partial gas pressure is not equal to one atmosphere. The general form of the equation is

$$E = E° - \frac{0.0591}{n} \log_{10} \frac{(conc\ C)^c (conc\ D)^d}{(conc\ A)^a (conc\ B)^b}$$

for the general reaction $aA + bB \rightarrow cC + dD$. The Nernst equation can be applied to determine the concentrations of ions in solution from voltage measurements. These concentrations can then be used to calculate values of equilibrium constants such as K_{sp}, K_a, K_b, and K_d. Another use of the Nernst equation is to calculate the potential of a half-cell. This is illustrated in Example 23.9 in the text.

Exercise 23.1
1. Given the standard reduction potentials for the following half-reactions:

Half-reaction	E°(volts)
$A^+(aq) + 1e^- \rightarrow A(s)$	-2.8
$B^{2+}(aq) + 2e^- \rightarrow B(s)$	-1.9
$C^{3+}(aq) + 3e^- \rightarrow C(s)$	-1.1
$D(l) + 2e^- \rightarrow D^{2-}(aq)$	$+2.4$

a. Select the species that is the best oxidizing agent; the best reducing agent.
b. Write the net ionic equation, calculate the standard voltage, and predict whether or not a spontaneous reaction will occur between B(s) and C^{3+}(aq).
c. Select two half-reactions which, when placed together, will form a cell that will deliver the highest possible voltage. Write a balanced equation for the reaction and calculate the cell voltage.
d. Calculate the value of $\Delta G°$ for the reaction in problem C.
e. Calculate the value of K for the reaction in problem C.

2. A half-cell contains a zinc electrode dipping into a 1.0 M solution of $Zn(NO_3)_2$. The other half-cell contains a lead electrode in a solution that is 1.0 M in Cl^- and saturated with lead chloride. Under these conditions the voltage is 0.49 V. Calculate the solubility product of $PbCl_2$.

Exercise 23.1 Answers
1. a. The best oxidizing agent (most easily reduced) is located in the lower left-hand column. The best reducing agent (most easily oxidized) is located in the upper right-hand column. Species D(l) is the best oxidizing agent, species A(aq) is the best reducing agent.
   b. The two half-reactions are:

   oxidation: $3 (B(s) \rightarrow B^{2+}(aq) + 2e^-)$; SOP = +1.9V

   reduction: $\underline{2(C^{3+}(aq) + 3e^- \rightarrow C(s))}$; SRP = -1.1V

   $3 B(s) + 2C^{3+}(aq) \rightarrow 3B^{2+}(aq) + 2C(s)$;   E° = +0.8V

   Notice that the E° is positive and that this reaction would be spontaneous. Further, the procedure of multiplying the oxidation half-reaction by 3 and the reduction reaction by 2 in the process of balancing the equation does not affect the potentials. A standard potential is independent of the number of electrons transferred in the reaction.

UNIT 23–OXIDATION-REDUCTION REACTIONS; SPONTANEITY AND EXTENT • 275

c. The combination of the best oxidizing agent and best reducing agent selected in problem A will be the pair that produces the greatest voltage. The equation and cell voltage may be obtained as follows:

oxidation: $\quad 2(A(s) \rightarrow A^+(aq) + 1e^-); \qquad\qquad SOP = +2.8\ V$

reduction: $\underline{\quad D(l) + 2e^- \rightarrow D^{2-}(aq) \qquad ; \qquad\qquad SRP = +2.4\ V\quad}$

$\qquad\qquad 2A(s) + D(l) \rightarrow 2A^+(aq) + D^{2-}(aq); \qquad E° = 5.2\ V$

d. The cell voltage has a high positive value, which indicates a spontaneous reaction; ΔG should be negative. The relationship $\Delta G° = -nFE°$ may be used. Two moles of electrons are transferred in the reaction.

$\qquad \Delta G° = -nFE° = (2)(-23.06)(5.2)\ \text{kcal} = -240\ \text{kcal}$

e. $\log_{10} K = \dfrac{nE°}{0.0591} = \dfrac{(2)(5.2)}{0.0591} = 176;\ K = 1 \times 10^{176}$

The extremely large value of K indicates that the reaction proceeds to completion. This is consistent with the large positive value of E° and the very negative value of $\Delta G°$.

2. First calculate E°.

$\qquad\qquad Zn(s) \rightarrow Zn^{2+}(aq) + 2e^-\ ; \qquad\qquad +0.76\ V$

$\underline{Pb^{2+}(aq) + 2e^- \rightarrow Pb(s) \qquad\qquad\quad ; \qquad\qquad -0.13\ V\quad}$

$Zn(s) + Pb^{2+}(aq) \rightarrow Zn^{2+}(aq) + Pb(s); \qquad\qquad E° = 0.63\ V$

Then use the Nernst equation.

$E = E° - \dfrac{0.0591}{2} \log_{10} \dfrac{\text{conc } Zn^{2+}}{\text{conc } Pb^{2+}}$

$0.49 = 0.63 - \dfrac{0.0591}{2} \log_{10} \dfrac{1}{\text{conc } Pb^{2+}}$

$\log_{10} \dfrac{1}{\text{conc } Pb^{2+}} = \dfrac{(0.49 - 0.63)(2)}{-0.0591} = 4.74 = 4 + 0.74$

$\dfrac{1}{\text{conc } Pb^{2+}} = 5.5 \times 10^4;\ \text{conc } Pb^{2+} = \dfrac{1}{5.5 \times 10^4} = 1.8 \times 10^{-5}$

$$K_{sp} = [Pb^{2+}][Cl^-]^2 = (1.8 \times 10^{-5})(1)^2 = 1.8 \times 10^{-5}$$

2. Reactions Involving Strong Oxidizing Agents

Highly electronegative nonmetals such as F_2, Cl_2, O_2, and oxyanions, in which the central atom is in a high oxidation state (MnO_4^-, ClO_3^-, $Cr_2O_7^{2-}$), are classified as strong oxidizng agents. These species have a tendency to gain electrons, as indicated by their high positive standard reduction potentials. Strong oxidizing agents may be used to synthesize compounds, to react with insoluble species to produce soluble compounds, and to analyze for a variety of ions and substances in analytical chemistry. Reactions involving strong oxidizing agents are discussed in Sections 23.4, 23.5, and 23.6 in the text.

A study of the chemical behavior of iron and dissolved oxygen in the redox process called corrosion has led to several different methods of protecting iron and steel materials from corrosion. Examples of these methods are cathodic protection, protective coatings, and inhibitors.

Exercise 23.2

1. A solution of $SnCl_2$ is used as a reagent in qualitative analysis to test for the presence of Hg^{2+} ion. Explain, using balanced equations, why solid tin is added to the $SnCl_2$ solution.
2. Explain why a tin can (steel that is tin plated) rusts rapidly when punctured, while a piece of galvanized iron (zinc plated) does not.
3. Potassium dichromate ($K_2Cr_2O_7$) reacts with Fe^{2+} in acid solution to produce Cr^{3+} and Fe^{3+}.
   a. Write a balanced equation for the reaction.
   b. It requires 28.0ml of 0.120 M $K_2Cr_2O_7$ solution to titrate 25.0ml of Fe^{2+} solution. Calculate the molarity of the Fe^{2+} solution.

Exercise 23.2 Answers

1. The $Sn^{2+}(aq)$ reacts with $Hg^{2+}(aq)$ as indicated in Equation 23.28 in the text. However, upon standing, the Sn^{2+} is oxidized to Sn^{4+} by oxygen. The solid tin is added to reduce the Sn^{4+} that forms. The equation for this reaction is

$$Sn(aq)^{4+} + Sn(s) \rightarrow 2Sn^{2+}(aq); E° = 0.29 \text{ V}$$

2. The Sn-Fe sets up a voltaic cell in which the Fe is the anode. In the Zn-Fe cell, Zn is the anode. Remember, the anode is where oxidation occurs.

UNIT 23–OXIDATION-REDUCTION REACTIONS; SPONTANEITY AND EXTENT • 277

3. a. $Cr_2O_7^{2-}(aq) + 14H^+(aq) + 6e^- \rightarrow 2Cr^{3+}(aq) + 7H_2O$

$\underline{\hspace{4em} 6(Fe^{2+}(aq) \rightarrow Fe^{3+}(aq) + e^-) \hspace{4em}}$

$Cr_2O_7^{2-}(aq) + 14H^+(aq) + 6Fe^{2+}(aq) \rightarrow 2Cr^{3+}(aq) + 6Fe^{3+}(aq) + 7H_2O$

b. $\dfrac{0.120 \text{ mole } Cr_2O_7^{2-}}{\text{liter}} \times 0.0280 \text{ liter} \times \dfrac{6 \text{ moles } Fe^{2+}}{1 \text{ mole } Cr_2O_7^{2-}} \times$

$\dfrac{1}{0.0250 \text{ liter}} = 0.806 \text{ M}$

SUGGESTED ASSIGNMENT XXIII

Text: Chapter 23

Problems: Numbers 23.1–23.5, 23.24, 23.25, 23.28, 23.31, 23.35, 23.36, 23.39, 23.41

Solutions to Assigned Problems Set XXIII

23.1

Find the standard potentials in Table 23.1 for the reduction and the oxidation half-reactions. Add the potentials with due regard to the signs.

reduction: $Cl_2(g) + 2e^- \rightarrow 2Cl^-(aq)$; SRP = +1.36V

oxidation: $\underline{\quad Sn^{2+}(aq) \rightarrow Sn^{4+}(aq) + 2e^-;\quad\quad\quad\quad}$ SOP = −0.15V

$Cl_2(g) + Sn^{2+}(aq) \rightarrow Sn^{4+}(aq) + 2Cl^-(aq)$; $E° = +1.21V$

Notice that the standard oxidation potential for the Sn^{2+}-Sn^{4+} reaction is equal in magnitude but opposite in sign to that of its reduction potential. The net cell potential is positive. This indicates that the reaction is spontaneous.

23.2

We need to refer to Table 23.1 to determine whether each of these species, when coupled with the O_2 reduction reaction, will produce a positive voltage. If the cell voltage is positive, the reaction will occur.

reduction: $O_2(g) + 4H^+(aq) + 4e^- \rightarrow 2H_2O$; SRP = +1.23V

oxidation: $\quad\quad\quad\quad 2Cl^-(aq) \rightarrow Cl_2(g) + 2e^-$; SOP = −1.36V

$E° = $ SRP + SOP = −0.13V. The O_2 will not oxidize Cl^- at these conditions.

oxidation: $Fe^{2+}(aq) \rightarrow Fe^{3+}(aq) + e^-$; SOP = −0.77V

$E° = $ SRP + SOP = +0.46V. The O_2 will oxidize Fe^{2+} at these conditions.

oxidation: $Co^{2+}(aq) \rightarrow Co^{3+}(aq) + e^-$; SOP = $-1.82V$

$E° = SRP + SOP = -0.59V$. The O_2 will not oxidize Co^{2+} at these conditions.

Notice that the strong oxidizing agents are listed near the bottom of the left column in Table 23.1; the strong reducing agents near the top of the right column. Generalizing, any species in the left column will oxidize any species above it in the right column.

23.3

The value of $E°$ is $+1.21V$ as determined in Problem 23.1. Using Equation 23.2, and noting that $n = 2$,

$$\Delta G° = -nFE° = -23.06 (2) (+1.21) \text{ kcal} = -55.8 \text{ kcal}$$

We use Equation 23.4 to relate the value of $E°$ to K. Again, $n = 2$.

$$\log_{10} K = \frac{nE°}{0.0591} = \frac{(2)(1.21)}{0.0591} = 40.9 = 0.9 + 40; K = 8 \times 10^{40}$$

Note that we predicted that the reaction in Problem 23.1 would be spontaneous at standard conditions since $E°$ is $+1.21V$. The value of $G°$ was very negative; this also indicates a spontaneous reaction. Finally, K was calculated and found to be a large positive value. This value of K indicates that the reaction proceeds far to the right, as written.

23.4

a. In this problem we determine the effect of concentrations on the cell voltage by use of the Nernst equation. Using the data for these conditions, substitute into the generalized form of the Nernst equation, Equation 23.6.

$$E = E° - \frac{0.0591}{n} \log_{10} \frac{(\text{conc } Sn^{4+})(\text{conc } Cl^-)^2}{(\text{conc } Sn^{2+})(\text{conc } Cl_2)}$$

$$= 1.21 - \frac{0.0591}{2} \log_{10} \frac{(0.10)(0.10)^2}{(0.10)(1)}$$

$$= 1.21 - \frac{0.0591}{2} \log_{10} 10^{-2} = 1.21 - \frac{0.0591}{2} (-2) = 1.27V$$

b. Using the Nernst equation in the above form,

$$1.09 = 1.21 - \frac{0.0591}{2} \log_{10} \frac{1}{(\text{conc Sn}^{2+})} = 1.21 + \frac{0.0591}{2} \log_{10} (\text{conc Sn}^{2+})$$

$$\log (\text{conc Sn}^{2+}) = \frac{(1.09 - 1.21)^2}{0.0591} = -4.1 = 0.9 - 5.0$$

Conc $Sn^{2+} = 8 \times 10^{-5}$

23.5

We use the concentration and volume of $KMnO_4$ to calculate the number of moles of $KMnO_4$ that react. We then use the coefficients in the balanced equation as conversion factors to determine the number of moles of I^- in 10.0 ml of I^- solution. Dividing the number of moles of I^- by 0.010 liter gives the concentration of I^-.

$$0.0250 \text{ liter} \times \frac{0.200 \text{ mole KMnO}_4}{\text{liter}} \times \frac{5 \text{ moles I}^-}{\text{mole KMnO}_4} \times \frac{1}{0.010 \ell} = 2.50$$

$$\frac{\text{mole I}^-}{\text{liter}} = 2.50 \text{ M}$$

23.24

As in Problem 23.1, split the reaction into oxidation and reduction half-reactions, obtain the standard potentials from Table 23.1, and add.

a. oxidation: $\qquad 2I^-(aq) \rightarrow I_2(s) + 2e^-$

reduction: $\underline{MnO_2(s) + 4H^+(aq) + 2e^- \rightarrow Mn^{2+}(aq) + 2H_2O}$

$\qquad MnO_2(s) + 4H^+(aq) + 2I^-(aq) \rightarrow Mn^{2+}(aq) + 2H_2O + I_2(s)$

SOP = -0.53V

SRP = $+1.23$V

E° = $+0.70$V

b. Note that the standard potentials must be obtained from the list of reactions in basic solution.

reduction: $S(s) + 2e^- \rightarrow S^{2-}(aq)$;   SRP = −0.48

oxidation: $\underline{H_2(g) + 2OH^-(aq) \rightarrow 2H_2O + 2e^-}$;   SOP = +0.83

$S(s) + H_2(g) + 2OH^-(aq) \rightarrow S^{2-}(aq) + 2H_2O$;   E° = +0.35

c. reduction: $AuCl_4^-(aq) + 3e^- \rightarrow Au(s) + 4Cl^-(aq)$

oxidation: $\underline{3(Ag(s) \rightarrow Ag^+(aq) + e^-)}$

$AuCl_4^-(aq) + 3Ag(s) \rightarrow Au(s) + 4Cl^-(aq) + 3Ag^+(aq)$

SRP = +1.00

SOP = −0.80

E = +0.20

Note that we do not multiply the oxidation potential by 3 even though the equation is multiplied by 3. The standard potential is independent of the number of electrons transferred. Notice also that $AuCl_4^-$ is a stronger oxidizing agent than Ag^+. Therefore, $AuCl_4^-$ is reduced and the Ag(s) is oxidized. The Ag electrode is the anode and Au the cathode in this cell; that is, Au plates out and Ag goes into solution.

23.25

In this problem, we are asked which of these nonspontaneous electrolytic cell reactions will occur when a maximum voltage of 1.5V is applied. We need to calculate the E° for each reaction and compare this value to 1.5V.

a. E° = SRP Cd^{2+} + SOP Cl^- = −0.40V + (−1.36V) = −1.76V

The reaction could not be accomplished with the application of 1.5 volts. At least 1.76V must be applied.

b. reduction: $2(2H_2O + 2e^- \rightarrow H_2(g) + 2OH^-(aq))$;   SRP = −0.83V oxidation: $2H_2O \rightarrow O_2(g) + 4H^+ + 4e^-$;   SOP = -1.23V

$$2H_2O \rightarrow 2H_2(g) + O_2(g); \qquad E° = -2.06 \text{ V}$$

At least 2.06V would have to be applied to bring about the electrolysis of water.

23.28

Using Table 23.1, we note that the more positive the potential the stronger the oxidizing agent. The order is

$Zn^{2+} < AuCl_4^- < MnO_4^- < H_2O_2$

23.31

$$K = \frac{[A^{2+}]^3}{[B^{3+}]^2} = \frac{(5.0 \times 10^{-3})^3}{(2.0 \times 10^{-2})^2} = 3.1 \times 10^{-4}$$

Equation 23.4 relates E° and K. Note that six moles of electrons are transferred in this reaction.

$$\log_{10} K = \frac{nE°}{0.0591} = \log_{10}(3.1 \times 10^{-4}) = \frac{6(E°)}{0.0591} ; \log_{10}(3.1 \times 10^{-4})$$

$= .49 + (-4) = -3.51$

$$E° = \frac{(-3.51)(0.0591)}{6} = -0.035\text{V}$$

$\Delta G° = -nFE° = -23.06(6)(-0.035) = +4.8 \text{ kcal}$

23.35

a. Calculate E° for the reaction.

$E° = \text{SRP } NO_3^- + \text{SOP Ag} = +0.96 - 0.80 = 0.16\text{V}$

Using the Nernst equation where n = 3,

$$E = E° - \frac{0.0591}{n} \log_{10} \frac{(\text{conc Ag}^+)^3 (\text{conc NO})}{(\text{conc } NO_3^-)(\text{conc } H^+)^4}$$

$$E = 0.16 - \frac{0.0591}{3} \log_{10} \frac{(0.10)^3 (1)}{(10)(10)^4}$$

$$E = 0.16 - \frac{0.0591}{3} \log_{10} (1 \times 10^{-8}) = 0.16 - \frac{0.0591(-8)}{3} = 0.16 +$$

$$0.16 = 0.32 \text{V}$$

b. This problem can be solved in a straightforward fashion if we realize that $E°$ for the cell is zero since the same electrode is involved in each half-cell. The expression is

$$E = 0.00 - \frac{0.0591}{1} \log_{10} \frac{(1)}{(0.001)} = 0.000 - 0.0591(3) = -0.18\text{V}$$

23.36

We use the Nernst Equation to calculate the concentrations of Au^{3+}:

$$1.00 = 1.50 - \frac{0.0591}{3} \log_{10} \frac{1}{(\text{conc Au}^{3+})}$$

$$= 1.50 + \frac{0.0591}{3} \log_{10} (\text{conc Au}^{3+})$$

$$\log_{10} (\text{conc Au}^{3+}) = \frac{(1.00 - 1.50)3}{0.0591} = -25; \text{ conc Au}^{3+} = 10^{-25}$$

$$K = \frac{[Cl^-]^4 \times [Au^{3+}]}{[AuCl_4^-]} = \frac{1^4 \times 10^{-25}}{1} = 10^{-25}$$

23.39

As in Chapter 22, write half-equations, multiply to conserve electrons, and add.

a. $2(e^- + NO_3^-(aq) + 2H^+(aq) \rightarrow NO_2(g) + H_2O)$

$\underline{\qquad NiS(s) \qquad\qquad\qquad \rightarrow Ni^{2+}(aq) + S(s) + 2e^- \qquad}$

$2NO_3^-(aq) + 4H^+(aq) + NiS(s) \rightarrow 2NO_2(g) + 2H_2O + Ni^{2+}(aq) + S(s)$

b. $O_2(g) + 4H^+(aq) + 4e^- \rightarrow 2H_2O$

$\underline{ 2(2I^-(aq) \rightarrow I_2(s) + 2e^-) }$

$O_2(g) + 4H^+(aq) + 4I^-(aq) \rightarrow 2H_2O + 2I_2(s)$

c. $Fe(s) \rightarrow Fe^{2+}(aq) + 2e^-$

d. The Sn^{2+} reduces the Hg^{2+} to Hg_2^{2+}. The Hg_2^{2+} reacts with Cl^- to produce $Hg_2Cl_2(s)$.

$2Hg^{2+}(aq) + Sn^{2+}(aq) + 2Cl^-(aq) \rightarrow Hg_2Cl_2(s) + Sn^{4+}(aq)$

23.41

a. $2MnO_4^-(aq) + 16H^+(aq) + 10I^-(aq) \rightarrow 2Mn^{2+}(aq) + 8H_2O + 5I_2(aq)$

b. Use conversion factors to relate the number of moles of $KMnO_4$ to the number of moles of I^- and divide by the volume of the sodium iodide solution.

$0.0189 \text{ liter} \times \dfrac{0.100 \text{ mole } MnO_4^-}{\text{liter}} \times \dfrac{10 \text{ moles } I^-}{2 \text{ moles } MnO_4^-} \times \dfrac{1}{0.0170 \text{ liter}}$

$= 0.556 \dfrac{\text{mole } I^-}{\text{liter}} = 0.556 M$

UNIT 24

Nuclear Reactions

The goal of this unit is to consider the characteristics of various types of nuclear reactions, their applications, and their implications.

INSTRUCTIONAL OBJECTIVES

Introduction
  Describe four ways in which nuclear reactions differ from ordinary chemical reactions.

24.1 Radioactivity
  1. Distinguish between natural and induced radioactivity.
  2. Describe the characteristics of alpha, beta, and gamma radiations.
  3. Write balanced nuclear equations when given:
     *a. The nuclide with its atomic number and mass number, and the type of radiation resulting from the decay. (Prob. 24.22a, b)
     *b. The nuclide and bombarding particle. (Probs. 24.1, 24.24)
     *c. The initial isotope (or isotopes) of a fission or fusion reaction and the symbol(s) of the element(s) after transmutation. (Prob. 24.22c, d)
  4. List the characteristics of a positron.
  5. Compare the applications of neutrons and positive ions in bombardment reactions.
  6. Briefly describe the basic principles of operation of instruments that detect and measure radiation such as a scintillation counter, cloud chamber, and Geiger-Muller counter.
  7. Recognize that the effect of ionizing radiation on living tissue is dependent on the energy and the type of radiation.
  8. Describe some applications of the use of radioactive isotopes in medicine, industry, and chemical research.

24.2 Rate of Radioactive Decay

*1. Use the first order rate law or the expression for the half-life to determine the rate constant, the half-life, or the amount of radioactive material left after a given time, including applications to estimate the age of an object. (Probs. 24.2, 24.25, 24.28)

24.3 Mass-Energy Relations

*1. Calculate Δm from Table 24.3; use the Einstein equation to relate Δm to ΔE. (Probs. 24.3, 24.30)

*2. Relate the mass decrement to the binding energy; interpret the binding energy per gram as a measure of the relative stabilities of nuclei. (Prob. 24.34)

24.4 Nuclear Fission

1. Recognize that the fission process produces a variety of radioactive species.
2. Compare the relative energies of fission reactions to ordinary chemical reactions.
3. Briefly discuss the advantages and disadvantages of nuclear power reactors.
4. Describe the principles of a breeder reactor.

24.5 Nuclear Fusion

1. Compare the energy produced from fusion reactions to that from fission reaction.
2. Explain why fusion reactions have very high activation energies.

SUMMARY

Nuclear reactions differ from ordinary chemical reactions in several important ways. These differences are summarized in the text in the introduction to this chapter. Nuclear transmutations occur spontaneously in nature; alternatively, they can be induced when target nuclei are bombarded with various kinds of particles. These reactions can be represented by nuclear equations. Balancing these equations is a straightforward process when the mass numbers and nuclear charges are written in the proper notation. The mass and nuclear charge characteristics of alpha, beta, gamma, the neutron, proton, and positron should be learned. The balancing process involves obtaining a sum of the mass numbers on the left of the equation equal to the sum on the right. Also, the sum of the nuclear charges on the left must equal that on the right.

Radioactive decay is a first order rate process that is independent of the temperature. Decay rates are often expressed in terms of the half-life, the time required for one-half of the sample to decay. The age of geological and archeological materials can be estimated based upon the rate of the radioactive decay process.

The action of radiation on living tissue produces ions which can result in mutations, the destructions of cells, or even the death of the organism. The effect of the radiation is related to the energy and the type of radiation. Applications of radioactive isotopes in medicine, industry, and chemical research require the use of instruments to detect and measure radioactivity.

Nuclear reactions result in a small difference in mass between the products and reactants. The Einstein equation relates the change in mass to the change in energy, $\Delta E = \Delta mc^2$. The binding energy of a nucleus is related to the mass decrement and is a measure of the stability of the nucleus. A comparison of the binding energy per gram of nuclei indicates the relative stabilities of different nuclei. The plot in Figure 24.5 shows that very light and very heavy nuclei are relatively unstable.

Certain heavy, unstable nuclei can split into smaller fragments in a process known as fission. A particular kind of nucleus can produce a variety of smaller nuclei, many of which are radioactive. A fission chain reaction occurs because more neutrons are produced than are consumed in the fission process.

The large amount of energy that is released in a fission reaction is related to a decrease in mass. The controlled fission process in a nuclear reactor is becoming increasingly important as a source of energy. The breeder reaction holds promise as a method of generating electrical energy in the future; however, numerous technological and safety problems must be solved before this process can be used on a large scale. Due to the very high activation energy of the fusion process, it appears that this will not be a practical method for the production of electrical energy in the near future.

Exercise 24.1
1. Complete the following nuclear equations:
    a. $^{209}_{82}Pb \rightarrow ^{209}_{83}Bi +$ _____
    b. $^{144}_{60}Nd \rightarrow ^{4}_{2}He +$ _____
    c. $^{63}_{29}Cu + ^{2}_{1}H \rightarrow 2^{1}_{0}n +$ _____
2. The rate constant for the decay of $^{197}_{78}Pt$ is 0.0385 hr^{-1}. Calculate:
    a. The number of grams of $^{197}_{78}Pt$ left after 12.0 hrs, starting with a sample weighing 0.106g.
    b. The half-life of this isotope.
3. A sample of uranium ore contains 1.240g of U-238 for every 0.2040g of Pb-206. Calculate the age of the ore. The half-life of U-238 is 4.50 × 10^9 years.

4. Use Table 24.3 and the relationship 2.15×10^{10} kcal = 1g to calculate the mass decrement (g/mole), the binding energy in kcal/mole, and in kcal/g for O-16.

Exercise 24.1 Answers
1. a. Notice that the mass numbers do not change; the nuclear charge increases from 82 to 83. The particle must have a mass number of zero and a nuclear charge of −1. This is a beta particle, $_{-1}^{0}e$.
   b. The mass number of the isotope must be 144 − 4 = 140. The nuclear charge is 60 − 2 = 58. Referring to the Periodic Table, element number 58 is cerium, symbol Ce. The isotope is $_{58}^{140}Ce$.
   c. Balancing mass numbers: 63 + 2 − 2 = 63
      Balancing nuclear charge: 29 + 1 − 0 = 30
      Element number 30 is zinc, the product is $_{30}^{63}Zn$.

2. a. $\log_{10} \dfrac{0.106}{x} = \dfrac{(0.0385)(12.0)}{2.30} = 0.201; \dfrac{0.106}{x} = $ antilog 0.201
      $= 1.59$

   $x = \dfrac{0.106}{1.59} = 0.0667g$

   b. $t_{½} = \dfrac{0.693}{k} = \dfrac{0.693}{0.0385} = 18.0$ hr

3. We must assume that all of the lead results from the decay of U-238. First calculate the number of grams of U-238 that must decay to produce 0.2040g of Pb-206.

   $0.2040g\ Pb \times \dfrac{238g\ U}{206g\ Pb} = 0.2357g\ U$

   If 0.2357g U-238 decayed and 1.240g remain, the initial amount of U-238 was 0.2357g + 1.240g = 1.476g

   Calculate the first order rate constant and then use Equation 24.11 to calculate the elapsed time.

   $k = \dfrac{0.693}{4.50 \times 10^9} = 1.54 \times 10^{-10}$

   $\log_{10} \dfrac{X_0}{X} = \dfrac{kt}{2.30}; \log_{10} \dfrac{1.476}{1.240} = \dfrac{(1.54 \times 10^{-10})t}{2.30} = 0.0757$

$$t = \frac{(0.0757)(2.30)}{1.54 \times 10^{-10}} = 1.13 \times 10^9 \text{ yrs}$$

4. $8\,^1_1H + 8\,^1_0n \rightarrow \,^{16}_8O$

$\Delta m = 15.99052g - 8(1.00728g) - 8(1.00867g) = -0.1371g$

The decrease in mass is 0.1371g when one mole of oxygen-16 is formed from 8 moles of protons and 8 moles of neutrons. The binding energy per mole can be calculated using the given conversion factor.

$$0.1371 \frac{g}{mole} \times (2.15 \times 10^{10}) \frac{kcal}{g} = 2.95 \times 10^9 \frac{kcal}{mole}$$

This means that 2.95×10^9 kcal would be evolved if 8 moles of protons and 8 moles of neutrons combined to form 1 mole of O-16.

The molar mass of O-16 is 15.99g. Using conversion factors,

$$2.95 \times 10^9 \frac{kcal}{mole} \times \frac{1 \text{ mole}}{15.99g} = 1.84 \times 10^8 \frac{kcal}{g}$$

Notice that the binding energy of 1.84×10^8 kcal/g for O-16 lies between the binding energies per gram of nuclei for C-12 and Al-27 in Table 24.4.

SUGGESTED ASSIGNMENT XXIV

Text: Chapter 24

Problems: Numbers 24.1–24.3, 24.22, 24.24, 24.25, 24.28, 24.30, 24.32, 24.34

Solutions to Assigned Problems Set XXIV

24.1

The total of the mass numbers and the total of the nuclear charges must be the same on both sides of the equation. The mass number of the isotope must be 96 + 4 − 1 = 99. Its nuclear charge must be 44 + 2 − 0 = 46. From the Periodic Table, we see that the product is an isotope of palladium, $^{99}_{46}Pd$:

$$^{96}_{44}Ru + ^{4}_{2}He \rightarrow ^{99}_{46}Pd + ^{1}_{0}n$$

24.2

a. The relationship to use for a first order reaction is:

$$k = \frac{0.693}{t_{1/2}} = \frac{0.693}{24.5 \text{ days}} = 2.83 \times 10^{-2} \text{ day}^{-1}$$

b. $\log_{10} \frac{X_0}{X} = \frac{kt}{2.30} = \frac{(2.83 \times 10^{-2} \text{ day}^{-1})(60 \text{ days})}{2.30} = 0.738$

Taking antilogs: $\frac{X_0}{X} = 5.47$. The fraction remaining is

$$\frac{X}{X_0} = \frac{1}{5.47} = 0.183$$

24.3

We use Table 24.3 to calculate Δm for the reaction. For one mole of Po:

Δm = mass of 1 mole $^{206}_{82}Pb$ + mass of 1 mole $^{4}_{2}He$ − mass of 1 mole $^{210}_{84}Po$ = 205.9295g + 4.00150g − 209.9368g = −0.0058g

Use Equation 24.16 to calculate ΔE, and then convert from ΔE for 1 mole to ΔE for 1 gram.

$$\Delta E = 2.15 \times 10^{10} \times \Delta m = 2.15 \times 10^{10} \times (-0.0058) = -1.25 \times 10^8 \text{ kcal};$$

$$-1.25 \times 10^8 \frac{\text{kcal}}{\text{mole}} \times \frac{1 \text{ mole Po}}{210 \text{g}} = -5.95 \times 10^5 \text{ kcal/g}$$

24.22

We use the same principles as in Problem 24.1.

a. $^{237}_{93}\text{Np} \rightarrow ^{4}_{2}\text{He} + ^{233}_{91}\text{Pa}$

b. $^{85}_{39}\text{Y} \rightarrow ^{0}_{-1}\text{e} + ^{85}_{38}\text{Sr}$

c. $^{12}_{6}\text{C} + ^{12}_{6}\text{C} \rightarrow ^{23}_{11}\text{Na} + ^{1}_{1}\text{H}$

d. $^{239}_{94}\text{Pu} + ^{1}_{0}\text{n} \rightarrow ^{130}_{50}\text{Sn} + 4^{1}_{0}\text{n} + ^{106}_{44}\text{Ru}$

24.24

a. $^{96}_{42}\text{Mo} + ^{2}_{1}\text{H} \rightarrow ^{97}_{43}\text{Tc} + ^{1}_{0}\text{n}$

b. $^{209}_{83}\text{Bi} + ^{4}_{2}\text{He} \rightarrow ^{211}_{85}\text{At} + 2^{1}_{0}\text{n}$

c. $^{10}_{5}\text{B} + ^{1}_{0}\text{n} \rightarrow ^{11}_{5}\text{B} + ^{0}_{0}\gamma$

d. $^{45}_{21}\text{Sc} + ^{1}_{0}\text{n} \rightarrow ^{1}_{1}\text{H} + ^{45}_{20}\text{Ca}$

24.25

Use the first order rate law and half-life expression as in Problem 24.2.

$$k = \frac{0.693}{t_{1/2}} = \frac{0.693}{29 \text{ yrs}} = 2.4 \times 10^{-2} \text{ yr}^{-1}$$

$$\log_{10} \frac{X_0}{X} = \frac{kt}{2.30} = \frac{(0.024)(100)}{2.3} = 1.04$$

Taking antilogs $\dfrac{X_0}{X} = 11$

The fraction remaining is $\dfrac{X}{X_0} = \dfrac{1}{11} = 0.091$

24.28

The value of the specific rate constant was calculated to be

1.21×10^{-4} yr^{-1} in Example 24.4. Since $X = 0.560\ X_0$,

$$\log_{10} \dfrac{X_0}{X} = \dfrac{(1.21 \times 10^{-4})(t)}{2.30} = \log_{10} \dfrac{1.000}{0.560} = \log 1.78 = 0.250$$

$$t = \dfrac{(0.250)(2.30)}{1.21 \times 10^{-4}} = 4.75 \times 10^3 \text{ yr}$$

24.30

The equation for the decay is:

$$^{245}_{97}\text{Bk} \rightarrow {}^{4}_{2}\text{He} + {}^{241}_{95}\text{Am}$$

Use data from Table 24.3 to calculate Δm for one mole of Bk.

Δm = mass of 1 mole $^{241}_{95}$Am + mass of 1 mole $^{4}_{2}$He − mass of 1 mole $^{245}_{97}$Bk

= 241.0045g + 4.0015g − 245.0129g = −0.0069g

$\Delta E = 2.15 \times 10^{10} \times (-0.0069) = -1.48 \times 10^8$ kcal

$1.48 \times 10^8 \dfrac{\text{kcal}}{\text{mole}} \times \dfrac{1 \text{ mole Bk}}{245\text{g}} = 6.04 \times 10^5$ kcal/g released

24.32

The decay will be spontaneous if the reaction results in a decrease in mass. For Al-26,

$$^{26}_{13}\text{Al} \rightarrow {}^{26}_{12}\text{Mg} + {}^{0}_{1}\text{e}$$

(The mass of an electron is 0.000549 = mass of a positron)

Δm = mass 1 mole $^{26}_{12}$Mg + mass 1 mole $^{0}_{1}$e − mass 1 mole $^{26}_{13}$Al

= 25.97600g + 0.00055g − 25.97977g = −0.00322g; spontaneous.

For Si-28, $^{28}_{14}$Si → $^{28}_{13}$Al + $^{0}_{1}$e

m = 27.97477g + 0.00055g − 27.96924g = +0.00608g; nonspontaneous.

24.34

To calculate the mass decrement for B-10, consider the reaction: 5^{1}_{1}H + 5^{1}_{0}n → $^{10}_{5}$B

Δm = mass 1 mole $^{10}_{5}$B − 5 (mass 1 mole $^{1}_{1}$H) − 5 (mass 1 mole $^{1}_{0}$n)

= 10.01019g − 5(1.00728g) − 5(1.00867g) = −0.06956g

Binding energy (kcal/mole): $0.06956 \frac{g}{mole} \times 2.15 \times 10^{10} \frac{kcal}{g} = 1.49 \times 10^{9}$ kcal/mole

Binding energy (kcal/g): $1.49 \times 10^{9} \frac{kcal}{mole} \times \frac{1\ mole}{10.01\ g} = 1.49 \times 10^{8}$ kcal/g.

UNIT 25

Polymers: Natural and Synthetic

The goal of this unit is to consider the formation, structure, and properties of natural and synthetic polymers.

INSTRUCTIONAL OBJECTIVES

25.1 Synthetic Polymers: Addition Type
 1. Recognize that addition polymers are generally synthesized from alkenes or alkene derivatives.
 *2. Sketch a portion of a polymer, given the formula of the monomer and the type of structure (i.e., head-to-tail, head-to-head and tail-to-tail, or random). (Probs. 25.1, 25.27)
 *3. Given the structure of a polymer, derive the structure of the monomer. (Prob. 25.29)
 *4. Distinguish between isotactic, syndiotactic, and atactic configurations. Sketch a portion of a polymer molecule, given the formula of the monomer and the type of configuration.

25.2 Synthetic Polymers: Condensation Type
 *1. Distinguish between addition and condensation polymerization reactions. Predict whether polymerization will occur by addition or condensation, given the formula of a monomer. (Prob. 25.30)
 *2. Sketch a portion of a polyester, given the formulas of an alcohol and an acid. Sketch the structure of the alcohol and acid, given the structure of the polyester. (Probs. 25.2, 25.29)

*3. Sketch a portion of a polyamide, given the formulas of an amine and an acid. Sketch the structure of the acid and amine, given the structure of the polyamide.

25.3 Natural Polymers: Carbohydrates
*1. Identify the asymmetric carbon atoms in a small molecule; predict the number of possible stereoisomers. (Prob. 25.34)
2. Compare the linkage of the glucose units in starch to their linkage in cellulose. Contrast the characteristics of starch to those of cellulose. (Prob. 25.35)

25.4 Natural Polymers: Proteins
*1. Sketch the form in which an amino acid would exist, given its isoelectric point and the pH of the solution. (Prob. 25.37)
*2. Sketch a possible portion of a protein chain, given the formulas of the amino acids. Sketch the structures of the amino acids, given the structure of a section of the protein molecule. (Prob. 25.4)
*3. Determine the number of amino acid residues or the minimum possible molecular weight of a polypeptide, given the appropriate analysis data. (Probs. 25.38, 25.40)
4. Briefly describe the mechanism by which an enzyme functions.
5. Define or identify the following terms: free radical, silicone, vulcanization, disaccharide, polysaccharide, peptide bond, and optical activity.

SUMMARY

Polymers are compounds that contain large molecules formed from the chemical combination of small molecular units called monomers. A given polymer may contain identical monomer units or be composed of many different monomers. In general, synthetic polymers, such as polyvinyl chloride and the polyesters, are simple structures that contain only one or two monomer units, compared to complex natural polymers such as proteins, which are composed of many different monomer units.

1. Synthetic Polymers

Addition polymers are formed from monomers that contain multiple bonds. These monomers are usually alkenes or derivatives of alkenes. The usually exothermic polymerization reaction is typically initiated by adding a reactive species called a free radical, a substance containing an unpaired electron. Free radicals are generally formed from organic peroxides.

The polymer chains formed from ethylene derivatives may result in different structures, depending on the manner in which the monomer units combine. The structures may be head-to-tail, head-to-head and tail-to-tail, or random. Since the bonding in the main carbon chain is tetrahedral and the attached atoms lack free rotation, several configurations may result, depending on the orientation of the R groups. Natta catalysts can be used to control polymer configurations to produce pure isotactic and syndiotactic forms rather than the atactic structures in which the R groups are randomly arranged.

Rubber is a natural addition polymer of the monomer isoprene. Synthetic rubber, along with other synthetic polymers, can be produced with a variety of properties. The variation in properties of these polymers is due to different additives and is caused by control of the chain length, cross-linking, and chain-branching.

Condensation polymers are formed by the elimination of a small molecule, usually water, from monomers that contain two (or more) functional groups. Polyesters are produced by the reaction between alcohols and organic acids. Dacron, Mylar, and Kodel are examples of polyesters. Polyamides are condensation polymers produced by the reaction of amines with organic acids — nylon is an example. Silicones are condensation polymers in which silicon rather than carbon is present in the chain.

Exercise 25.1

1. Vinyl alcohol, $CH_2=CHOH$, forms a head-to-tail addition polymer.
   a. Sketch a portion of a polyvinyl alcohol molecule.
   b. Can you suggest why this polymer is one of the few polymers that have an appreciable solubility in water?
   c. The vinyl alcohol polymer can cross-link between chains to form a nonmelting polymer. Water is eliminated in this reaction. Sketch a portion of the network that results when two adjacent chains undergo cross-linking.

2. Draw the structures of the monomers used to make the following polymer.

$$-O-(CH_2)_3-O-\underset{\underset{O}{\|}}{C}-(CH_2)_2-\underset{\underset{O}{\|}}{C}-O-$$

Exercise 25.1 Answers

1. a. The monomer units are oriented in the same direction with an OH group on every other carbon atom.

$$\begin{array}{cccc} H & H & H & H \\ | & | & | & | \\ -C-&C-&C-&C- \\ | & | & | & | \\ OH & H & OH & H \end{array}$$

b. Due to the OH groups, there is hydrogen bonding with H_2O molecules.
c. Two hydroxide groups on adjacent chains interact to form an oxygen bridge with elimination of water.

$$\begin{array}{c}
HHHH \\
|||| \\
-C-C-C-C- \\
|||| \\
OHOH \\
|H|H \\
-C-C-C-C- \\
|||| \\
HHHH
\end{array}$$

2. This is a section of a polyester molecule. The polyester is formed from the condensation reaction between a dihydroxy alcohol and a dicarboxylic acid.

The alcohol is $HO-(CH_2)_3-OH$

The acid is $HO-\overset{\overset{O}{\|}}{C}-CH_2-CH_2-\overset{\overset{O}{\|}}{C}-OH$

2. Natural Polymers

Many chemical substances produced by living organisms are complex polymers containing several different kinds of monomer units. These natural polymers not only are essential to living organisms but are also used in the manufacture of paper, fibers for various kinds of fabrics, and as raw materials for the production of other polymers.

Carbohydrates are compounds of carbon, hydrogen, and oxygen with the general formula of $C_n(H_2O)_m$. Simple sugars or monosaccharides usually contain five or six carbon atoms with hydroxyl groups on each carbon atom except one, which contains a carbonyl group. When the carbonyl group is on a terminal carbon atom, as in glucose, the sugar is an aldehyde. Other sugars such as ribulose, with five carbon atoms, and fructose, with six carbon atoms, contain the ketone group.

Solutions of glucose and most other sugars rotate the plane of a transmitted beam of polarized light. This optical activity is ordinarily found in organic substances in which there are one or more carbon atoms bonded to four different groups; such carbon atoms are said to be asymmetric. The presence of an asymmetric carbon atom leads to the existence of two different isomers which are nonsuperimposable mirror images of each other, and which cannot be converted one to the other without breaking chemical bonds. Since there are four asymmetric carbon atoms in glucose (two different positions for the hydroxyl group on carbon atoms numbers 2, 3, 4, and 5), there are 2^4, or 16, possible stereoisomers.

Since the carbon chain is flexible, with free rotation around the carbon-carbon single bonds, most five and six carbon atom monosaccharides usually exist as three-dimensional ring structures rather than in the linear form.

Two monosaccharide units in ring form may condense, with the elimination of water, to form a disaccharide containing an oxygen bridge.

Maltose and sucrose are disaccharides, which, upon hydrolysis, yield glucose, and glucose and fructose, respectively. Polysaccharides contain chains of monosaccharides linked by oxygen bridges. Starch and cellulose may be considered to be condensation polymers of alpha and beta glucose units, respectively.

Proteins are polymeric substances composed of long chains of up to 20 different α-amino acid monomer units. The amino acids combine by a condensation reaction with the formation of a peptide bond and the elimination of water. These polypeptides have an amino group on one end of the chain and a carboxyl group at the other. There are many different proteins since there are up to 20 different monomer units in a polypeptide and many different ways the amino acid residues can be arranged. Protein structure determination involves establishing the relative number of moles of each amino acid as well as the sequence of these monomer units. The basis of many of the procedures used in the separation and identification of the monomer units is related to the charges, including those on the R groups, on the α-amino acids. Depending on the pH of the solution, α-amino acids can exist as either cations or anions. At a particular pH, the isoelectric point, an amino acid will be uncharged.

The conformation of a protein molecule is determined primarily by the interaction of the many groups on the same or adjacent molecules. If the R groups of the amino acid residues are small, hydrogen bonding is maximized when the chains form a sheet (Figure 25.9). When the R groups are bulky, as in many proteins, maximum hydrogen bonding occurs when the protein chain is in the form of a helical coil (Figure 25.10). The folded sheets or helical coils can be twisted to result in long fibrous structures or in a globular form as in cytochrome C (Figure 25.11). Enzymes are globular proteins that catalyze specific chemical reactions in living systems.

Exercise 25.2
1. Which of the following compounds could have optically active forms? For those which may be active, indicate the number of asymmetric carbon atoms present and the number of optically active stereoisomers that each compound could have.

a.
$$H-\underset{\underset{H}{|}}{\overset{\overset{H}{|}}{C}}-\underset{\underset{H}{|}}{\overset{\overset{H}{|}}{C}}-\underset{\underset{H}{|}}{\overset{\overset{OH}{|}}{C}}-C=O$$

b.
$$HO-\underset{\underset{H}{|}}{\overset{\overset{H}{|}}{C}}-\underset{\underset{H}{|}}{\overset{\overset{H}{|}}{C}}-\underset{\underset{H}{|}}{\overset{\overset{H}{|}}{C}}-\overset{\overset{O}{\|}}{C}-\underset{\underset{H}{|}}{\overset{\overset{H}{|}}{C}}-H$$

c.
$$O=\underset{\underset{OH}{|}}{C}-\underset{\underset{OH}{|}}{\overset{\overset{H}{|}}{C}}-\underset{\underset{OH}{|}}{\overset{\overset{H}{|}}{C}}-\underset{\underset{OH}{|}}{C}=O$$

2. The isoelectric points of isoleucine, glutamic acid, and glycine are 6.04, 3.08, and 6.06, respectively. Predict the sign of the charge, if any, on each of the following amino acid molecules in a solution for which the pH equals 2.

3. Write structural formulas for the four possible dipeptide molecules produced when valine, $(CH_3)_2 CHC-HCOOH$, and glycine,
$$\overset{|}{NH_2}$$
$H_2 NCH_2 COOH$, are mixed and allowed to react.

Exercise 25.2 Answers

1. Compound (a) could be optically active since it has one asymmetric carbon atom (the number two carbon atom). It could have two possible stereoisomers.

   Compound (b) would not be optically active. It contains no asymmetric carbon atoms.

   Compound (c) has two asymmetric carbon atoms (carbon atoms two and three). It would be optically active with four possible stereoisomers.

2. All three amino acids would be positively charged. The NH_2 groups would gain a proton and exist in the cation form.

3. Val-Val
$$\underset{\underset{CH_3}{\diagup}}{\overset{\overset{CH_3}{\diagdown}}{H}}-\underset{\underset{NH_2}{|}}{\overset{\overset{H}{|}}{C}}-\overset{\overset{O}{\|}}{C}-\underset{\underset{H}{|}}{N}-\underset{\underset{C=O}{|}}{\overset{\overset{CH_3\diagdown\overset{H}{\underset{|}{C}}\diagup CH_3}{|}}{C}}-H$$
$$|$$
$$OH$$

Gly-Gly

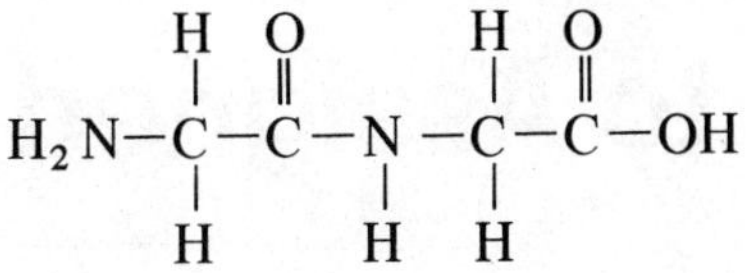

$$H_2N-\underset{\underset{H}{|}}{\overset{\overset{H}{|}}{C}}-\overset{\overset{O}{\|}}{C}-\underset{\underset{H}{|}}{N}-\underset{\underset{H}{|}}{\overset{\overset{H}{|}}{C}}-\overset{\overset{O}{\|}}{C}-OH$$

Val-Gly

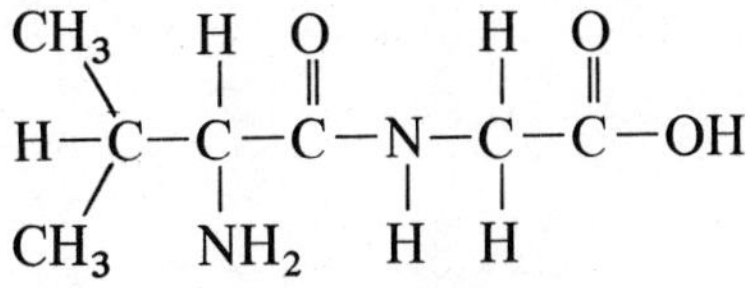

$$H-\underset{\underset{CH_3}{|}}{\overset{\overset{CH_3}{|}}{C}}-\underset{\underset{NH_2}{|}}{C}-\overset{\overset{O}{\|}}{C}-\underset{\underset{H}{|}}{N}-\underset{\underset{H}{|}}{C}-\overset{\overset{O}{\|}}{C}-OH$$

Gly-Val

$$H_2N-\underset{\underset{H}{|}}{\overset{\overset{H}{|}}{C}}-\overset{\overset{O}{\|}}{C}-\underset{\underset{H}{|}}{N}-\underset{\underset{C=O}{|}}{\overset{\overset{CH_3\ \ H\ \ CH_3}{\diagdown|\diagup}}{C}}-H$$

$$OH$$

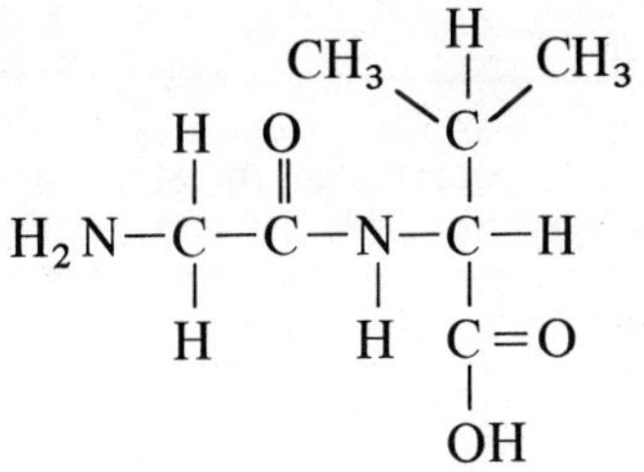

SUGGESTED ASSIGNMENT XXV

Text: Chapter 25

Problems: Numbers 25.1–25.4, 25.27, 25.29, 25.30, 25.34, 25.35, 25.37, 25.38, 25.40

Solutions to Assigned Problems Set XXV

25.1

a. The monomer units are oriented in the same direction in a head-to-tail structure.

$$-\underset{\underset{H}{|}}{\overset{\overset{H}{|}}{C}}-\underset{\underset{C_6H_5}{|}}{\overset{\overset{H}{|}}{C}}-\underset{\underset{H}{|}}{\overset{\overset{H}{|}}{C}}-\underset{\underset{C_6H_5}{|}}{\overset{\overset{H}{|}}{C}}-$$

b. One carbon-carbon double bond is broken and two carbon-carbon single bonds are formed each time a molecule of monomer is added to the chain. From Tables 4.2 and 8.5, B.E. C—C is 83 kcal/mole and B.E. C=C is 143 kcal/mole.

ΔH = B.E. C=C − 2B.E. C—C = 143 kcal − 2(83 kcal) = −23 kcal

25.2

Water is eliminated when this dihydroxy alcohol reacts with the dicarboxylic acid. The ester that is formed has groups at each end of the molecule that can react to produce a long-chain polymeric ester.

$$-O-CH_2-CH_2-O-\underset{\underset{O}{\|}}{C}-\underset{\underset{O}{\|}}{C}-O-$$

25.3

a. Weight per mole of $C_6H_{10}O_5$ unit = 6(12.0g) + 10(1.01g) + 5(16.0g) = 162g

UNIT 25—POLYMERS: NATURAL AND SYNTHETIC · 303

$$\%C = \frac{72.0g}{162g} \times 100 = 44.4\%$$

$$\%H = \frac{10.0g}{162g} \times 100 = 6.2\%$$

$$\%O = \frac{80.0g}{162g} \times 100 = 49.4\%$$

b. $(1.0 \times 10^4)(1.62 \times 10^2) = 1.62 \times 10^6$

25.4

The glycine molecules combine by the reaction of the acid group on one molecule with the basic group on another.

```
      H O      H O      H O      H O
      | ||     | ||     | ||     | ||
H—N—C—C—N—C—C—N—C—C—N—C—C—OH
  |  |    |  |    |  |    |  |
  H  H    H  H    H  H    H  H
```

25.27

In a head-to-head, tail-to-tail polymer, the successive monomer units are oriented in opposite ways, the CNs occur in pairs on adjacent carbon atoms.

```
   H  H  H  H  H  H  H  H  H  H
   |  |  |  |  |  |  |  |  |  |
—C—C—C—C—C—C—C—C—C—C—
   |  |  |  |  |  |  |  |  |  |
   H  CN CN H  H  CN CN H  H  CN
```

25.29

a. The acid is *p*-phthalic acid. HO—C(=O)—⟨benzene⟩—C(=O)—OH

  The alcohol is ethylene glycol. HO—CH₂—CH₂—OH

$$\text{HO}-\overset{\overset{\displaystyle O}{\|}}{C}-\!\!\!\bigcirc\!\!\!-\overset{\overset{\displaystyle O}{\|}}{C}-\text{OH}$$

$$\text{HO}-\overset{\overset{\displaystyle H}{|}}{\underset{\underset{\displaystyle H}{|}}{C}}-\overset{\overset{\displaystyle H}{|}}{\underset{\underset{\displaystyle H}{|}}{C}}-\text{OH}$$

b. Dichloroethylene.

$$\begin{array}{c} H \\ \diagdown \\ Cl \end{array} C=C \begin{array}{c} H \\ \diagup \\ Cl \end{array}$$

c. Lysine. $NH_2-\overset{\overset{H}{|}}{\underset{\underset{H}{|}}{C}}-\overset{\overset{H}{|}}{\underset{\underset{H}{|}}{C}}-\overset{\overset{H}{|}}{\underset{\underset{H}{|}}{C}}-\overset{\overset{H}{|}}{\underset{\underset{H}{|}}{C}}-\overset{\overset{H}{|}}{\underset{\underset{H}{|}}{C}}-\overset{\overset{O}{\|}}{C}-OH$

25.30

The polymer in (a) is a condensation polymer that results from the reaction between a dicarboxylic acid and a dihydroxy alcohol. The polymer in (b) is an addition polymer. The structure in (c) is a protein.

25.34

a. There are two asymmetric carbon atoms, carbon atoms numbers 2 and 3.

b. There are no asymmetric carbon atoms.

c. There are two asymmetric carbon atoms, carbon atoms numbers 2 and 3.

25.35

In cellulose, the oxygen bridge between the cyclic glucose units is in the β position. The long, essentially unbranched chains in cellulose allow strong hydrogen bonding between the polymer chains. This gives great strength to the cellulose fibers. The oxygen bridge in starch is in the α position and does not allow strong hydrogen bonding between chains.

25.37

$$\begin{array}{c} H \\ \diagdown \\ \diagup \\ H \end{array} N - \overset{\overset{H}{|}}{\underset{\underset{H}{|}}{C}}-\overset{\overset{H}{|}}{\underset{\underset{H}{|}}{C}}-\overset{\overset{H}{|}}{\underset{\underset{H}{|}}{C}}-\overset{\overset{H}{|}}{\underset{\underset{H}{|}}{C}}-\overset{\overset{H}{|}}{\underset{\underset{NH_2}{|}}{C}}-\overset{\overset{O}{\|}}{\underset{\underset{O^\ominus}{|}}{C}}$$

In basic solution, lysine loses a proton and exists as an anion. In acid solution, each $-NH_2$ may pick up a proton to produce a doubly charged cation.

$$\overset{\oplus}{} H-\underset{H}{\overset{H}{\underset{|}{\overset{|}{N}}}}-\underset{H}{\overset{H}{\underset{|}{\overset{|}{C}}}}-\underset{H}{\overset{H}{\underset{|}{\overset{|}{C}}}}-\underset{H}{\overset{H}{\underset{|}{\overset{|}{C}}}}-\underset{H}{\overset{H}{\underset{|}{\overset{|}{C}}}}-\underset{NH_3^{\oplus}}{\overset{H}{\underset{|}{\overset{|}{C}}}}-\overset{O}{\overset{\|}{C}}-OH$$

25.38

Using only one of each amino acid, there are six possible combinations. If any number of each amino acid is used (giving combinations such as Gly-Gly-Gly, or Gly-Gly-Val), there are 27 possible tripeptides.

25.40

$$\text{moles leucine} = \frac{1.65 \times 10^{-5} \text{g}}{131 \text{g/mole}} = 1.26 \times 10^{-7} \text{ mole}$$

$$\text{moles isoleucine} = \frac{2.48 \times 10^{-5} \text{g}}{131 \text{g/mole}} = 1.89 \times 10^{-7} \text{ mole}$$

$$\frac{1.87 \times 10^{-7} \text{ mole isoleucine}}{1.26 \times 10^{-7} \text{ mole leucine}} = 1.5 \text{ moles isoleucine per mole leucine}$$

There are 3 moles of isoleucine per 2 moles of leucine. Therefore, there must be a minimum of 2 moles of leucine per mole of enzyme.

1.26×10^{-7} mole leucine $\times \dfrac{1 \text{ mole enzyme}}{2 \text{ moles leucine}} = 6.30 \times 10^{-8}$ mole enzyme

$1.00 \times 10^{-3} \text{g} / 6.3 \times 10^{-8}$ mole $= 1.59 \times 10^4 \text{g/mole}$

The minimum possible molecular weight is 15,900.